BUILDING FOR SAFETY

COMPENDIUM

Building for Safety Compendium

An annotated bibliography and information
directory for safe building

Andrew Clayton and Ian Davis
The Oxford Centre for Disaster Studies

INTERMEDIATE TECHNOLOGY PUBLICATIONS 1994

Practical Action Publishing Ltd
25 Albert Street, Rugby, CV21 2SD, Warwickshire, UK
www.practicalactionpublishing.com

© Andrew Clayton, Ian Davis and The Oxford Centre for Disaster Studies 1994

First published 1994

ISBN 10: 1 85339 181 6
ISBN 13 Paperback: 9781853391811
ISBN Library Ebook: 9781780442105
Book DOI: https://doi.org/10.3362/9781780442105

All rights reserved. No part of this publication may be reprinted or reproduced or utilized in any form or by any electronic, mechanical, or other means, now known or hereafter invented, including photocopying and recording, or in any information storage or retrieval system, without the written permission of the publishers.

A catalogue record for this book is available from the British Library.

The authors, contributors and/or editors have asserted their rights under the Copyright Designs and Patents Act 1988 to be identified as authors of their respective contributions.

Since 1974, Practical Action Publishing has published and disseminated books and information in support of international development work throughout the world. Practical Action Publishing is a trading name of Practical Action Publishing Ltd (Company Reg. No. 01159018), the wholly owned publishing company of Practical Action. Practical Action Publishing trades only in support of its parent charity objectives and any profits are covenanted back to Practical Action (Charity Reg. No. 247257, Group VAT Registration No. 880 9924 76).

Reasonable efforts have been made to publish reliable data and information, but the author and publisher cannot assume responsibility for the validity of all materials or for the consequences of their use.

Typeset by Dorwyn, Rowlands Castle, Hants.

The manufacturer's authorised representative in the EU for product safety is Lightning Source France, 1 Av. Johannes Gutenberg, 78310 Maurepas, France.
compliance@lightningsource.fr

Contents

1. General disaster management with particular reference to shelter and housing provision 3
2. Training and communication 8
3. Disaster-resistant construction 11
4. Manuals and guidelines on construction 14
5. Building materials 16
6. Services for human settlements 19
7. Project planning and evaluation 20
8. Community organization 23
9. Gender issues 26
10. Financial management 28
11. Case studies 30
12. Audio-visual materials 33
13. General resource material 35
14. Journals and newsletters 37
15. Directory of organizations 40

Introduction

This annotated source compendium was compiled and written by Andrew Clayton and Ian Davis as part of the ODA-funded Building for Safety Project jointly undertaken by Oxford Centre for Disaster Studies and Cambridge Architectural Research. The other members of the Oxford team, Alistair Cory and Yasemin Aysan, provided support and suggestions for the compendium. It is intended to accompany the *Guidelines for Developing Building Improvement Programmes in Disaster Prone Areas* which has been written jointly by the Oxford project team.

The purpose of the compendium is to assist project workers and decision-makers involved in building improvement programmes by indicating what further resources are available to them. The compendium covers the major areas covered in the guidelines. Key texts in each area have been identified, and annotations provided. The holistic approach to building improvement programmes advocated in the guidelines, incorporating both technical and social considerations, is reflected in the broad range of materials presented in the compendium.

Given the practical focus of the *Building for Safety Compendium*, only material that is cheaply and easily available has been included, and the selection of material has been confirmed to practical books and manuals rather than academic, theoretical studies. No attempt has been made to produce a comprehensive bibliography. In order to assist those organizations that are unable to obtain the necessary foreign exchange for book purchase, a number of UNCHS(HABITAT) publications have been included in the compendium, since these can be ordered without charge.

A directory of organizations has also been included. These organizations have developed internationally recognized expertise in areas relevant to building improvement programmes, or have an important networking role.

It is hoped that users of the compendium will use this document as a basis for the development of their own source compendiums, which would incorporate locally produced written and visual materials, and their own directory of organizations operational in their country or region. Furthermore, given that most of the material in this compendium is written in English, there is a need to document locally produced material written in local or national languages. Suggested sources of information that would assist this process are provided below.

Each entry in this compendium provides the following information: title of publication or visual material; the name of the author or organization responsible for the publication; a review of the main contents of the publication and its relevance to building improvement programmes, together with an indication of which readership will find it useful; the name of the publisher; the year of publication; the number of pages to give some indication of size; the price, if known, although it should be noted that these may be subject to change; and an address of where the publication is available from.

Any enquiries concerning *Building for Safety Compendium* should be addressed to:

Andrew Clayton
Oxford Centre for Disaster Studies
PO Box 137
Oxford OX4 1BB.
UK.

How to obtain basic information to organize a building improvement programme

The following advice is primarily to assist a person who is unfamiliar with the subject but who has to gain certain information to assist in running a project. In many areas locally available knowledge and resources may be limited.

Initially, information is needed in five broad areas: 1. community-level training programmes; 2. building/planning of low-cost dwellings; 3. special safety factors that relate to local hazards; 4. governmental (both central and local) links that relate to local planning constraints, by-laws, grants and reconstruction policies; and 5. financial sources of grants/loans.

First, carefully read this compendium to see if there are sources that can be approached which relate to your programme. Then explore local sources in order to obtain information that is directly relevant to the specific needs of your particular programme. The following information sources are suggested to guide you in this.

1. Community-level training programmes

- Local NGOs that work with local communities (not necessarily in the building sector)
- Governmental offices concerned with community development and social services
- Local training institutes, and
- Local academic bodies, such as departments of education and social sciences.

2. Building/planning of low-cost dwellings

- Government departments of housing, planning or public works
- Local building contractors who normally build low-cost dwellings in the locality, and
- Local academic bodies, such as departments of structural engineering, planning and construction.

3. Special safety factors that relate to local hazards

- Local government officials that are involved in disaster planning and management
- Local academic bodies, such as departments of seismology/geography (earthquakes), hydrology (floods), meteorology (winds), geology (landslides)
- Local leaders who may be able to provide information on the effects of past disasters, and
- Regional disaster centres and regional offices of UN agencies.

4. Government and local government links

- Government officials responsible for reconstruction planning, and
- Local law enforcement officers in building/planning/public works/housing offices who are responsible for building codes and land use controls.

5. Financial sources

- Banks providing loans for housing
- Charitable foundations, and
- Local NGOs that fund development projects.

1. General disaster management with particular reference to shelter and housing provision

The intention of this section is to identify key documents that provide an overview of both pre-disaster planning and post-disaster response with particular reference to shelter needs. In order to clarify the main subject areas, this section has been divided according to the following headings: general disaster management; risk assessment; disaster mitigation; and post-disaster shelter.

General disaster management

Disasters and Development
Cuny, F.

This influential book examines the nature of disaster response from the perspective of long-term development. The author argues for a reassessment of disaster relief so that it supports and encourages the social and economic development of communites following disasters. It focuses primarily on disaster assistance following earthquakes and hurricanes and pays particular attention to post-disaster housing programmes. The book is highly recommended for the wealth of practical experience that the author brings to this key subject.

Oxford University Press; 1983; 278pp; £20
Available from: Oxford University Press, Walton St, Oxford, UK and Oxford University Press, New York, USA.

UNDP/DHA-UNDRO Disaster Management Training Programme – Training Materials

The UNDP/DHA-UNDRO Disaster Management Training Programme was set up to provide a fresh approach to disaster management training for developing countries. As part of the programme, a number of training modules have been produced. Each training module provides up-to-date technical information, clearly presented for a non-specialist readership. The DMTP has produced (or is in process of producing) the following training modules and trainer's guide (S – also in Spanish; F – also in French):

- *Overview of Disaster Management*
- *Disaster Preparedness* (S)
- *Disasters and Development* (S)
- *Disaster Assessment* (S)(F)
- *Disaster Mitigation* (S)
- *Displaced Persons in Civil Conflict* (S)
- *Drought and Famine*
- *Humanitarian Principles: operational dilemas*
- *Information Management and Communication*
- *International Law and Emergencies*
- *Introduction to Hazards*
- *Logistics*
- *Rehabilitation*
- *Vulnerability and Risk Assessment* (S)
- *Disaster Economics, and*
- *Disasters and the Environment*

Two sets of training guidelines have also been produced:

- *Guidelines for Trainers Leading Disaster Management Workshops*: part I and II
- *Guide to Developing Training Strategies and Programmes*

One copy of the training modules are available free on request. The trainers' guides are given free to individuals or organizations involved in training. (For full details of DMTP materials, write to address below).

Available from: University of Wisconsin, Disaster Management Center, Department of Engineering Professional Development, 432 North Lake St., Madison, WI 53706, USA.

Disaster Management: a disaster manager's handbook
Carter, Nick

Guidelines for all aspects of disaster management are provided in this handbook. It is aimed specifically for agencies and officials working in disaster-prone developing countries, especially in Asia and the Pacific region. It is the most comprehensive handbook to date and is highly recommended for its general overview of disaster management.

ADB; 1991; 417pp; free
Available from: Information Office, Asian Development Bank, PO Box 798, 1099 Manila, Philippines.

Christian Perspectives on Disaster Management
Davis, I. and Wall, M. (eds)

The material in this manual has been adapted from the texts of a wide range of documents on disaster management and is intended for the use of participants attending training workshops in disaster management. Although tailored specifically for Christian groups, the manual provides useful training material on developing both management skills and understanding of a wide range of disaster subjects including housing.

Interchurch Relief and Development Alliance; 1993; 269pp
Available from: Training Unit, Tearfund, 100 Church Rd, Middlesex, TW11 8QE, UK.
Also available in Portuguese.

Risk assessment

Rising From the Ashes: development strategies in times of disasters
Anderson, M. and Woodrow, P.

The aim of this book is to assist NGOs and others to develop approaches to the provision of emergency assistance in ways that support long-term development. This book is based on 30 case studies of emergency relief programmes that have promoted long-term development. It presents a framework for assessing the needs, vulnerabilities and capabilities of afflicted communities and argues that development efforts must be based on existing capacities. Part I of the book presents the lessons learned and guidelines derived from these case studies and through discussion and consultation with NGOs. Part II of the book presents 11 of these case studies in detail including three on reconstruction. When published this was the only available study to look at vulnerability and its link to underdevelopment, and to suggest useful ways to reduce it.

Westfield Press/UNESCO; 1989; 338pp
Available from: Westfield Press, 5500 Central Avenue, Boulder, Colorado 80301, USA or UNESCO, 7 Place de Fontenoy, 75700 Paris, France.

At Risk: natural hazards, people's vulnerability and disasters
Blaikie, P., Cannon, T., Davis, I. and Wisner, B.

An examination of patterns of vulnerability to all major natural hazards is the theme of this book. A model is developed to indicate the way vulnerability is created from root causes by way of pressures to unsafe conditions. The model is then reversed to indicate how vulnerability can be reduced. The book will be of interest to readers with an interest in building safety because of its many examples of how vulnerability has developed in the past, and of ways to reduce it through political, developmental and technical measures. There is an extensive bibliography drawn from the authors wide international experience of the subject.

Routledge; forthcoming 1994; 300pp approx
Available from: Routledge, 11 New Fetter Lane, London EC4P 4EE.

Natural Disasters
Alexander, D.

This book provides a comprehensive overview of the physical, technological and social aspects of natural disasters. A major concern of the book is with the interaction between geophysical agents and human vulnerability and response. It is aimed primarily towards an academic audience rather than for practioners.

UCL Press; 1993; 632; £19.95
Available from: UCL Press, University College London, Gower St, London, WC1E 6BT.

Disaster mitigation

Disaster Mitigation: a community-based approach
Maskrey, A.

The importance of this book is its treatment of disaster mitigation, including building improvement programmes, in the context of people's exposure to both natural hazards and social and economic factors. It is the first book to show how small NGOs can reduce risks from floods and earthquakes through mobilizing a local community to reduce their own risks by working to improve their buildings and infrastructure. The author draws extensively on his experience of working in Peru and on case study material from other Latin American countries. The book is divided into two parts: the first part addresses the problem of vulnerability to disaster, emphasizing the relationship between vulnerability and poverty; the second part of the book looks at approaches to mitigating disaster. The top-down approach to disaster mitigation, favoured by governments is criticized for addressing only the effects of hazards and not the underlying causes of vulnerability. Instead an approach is recommended in which disaster mitigation planning and implementation is carried out by community organizations. Not only does this encourage empowerment of the groups at risk, but also, the author argues, improves the effectiveness of the mitigation measures.

Oxfam; 1989; 100pp; £4.95
Available from: Oxfam, 274 Banbury Rd, Oxford OX2 7DZ, UK and Intermediate Technology Publications, 103/105 Southampton Row, London WC1B 4HH, UK. Also available in Spanish from Intermediate Technology Publications Peru, Casilla 18-0620, Lima 18, Peru.

Disaster Mitigation in Asia and the Pacific
Asian Development Bank

General analyses and strategies for disaster mitigation of natural disasters in the Asia and Pacific region are the subject of this comprehensive study. It is addressed primarily for officials from countries within this region to assist in policy-making. There is a technical background paper which covers natural hazards in the Asia-Pacific region, vulnerability analysis, mitigation methods, and implementation of mitigation. There then follows four detailed studies of disaster mitigation in Bangladesh, Nepal, the Philippines and the South Pacific Islands.

ADB; 1991; 392pp; free
Available from: Information Office, Asian Development Bank, PO Box 798, 1099 Manila, Philippines.

Mitigating Natural Disasters: phenomena, effects and options. A manual for policy-makers and planners
Office of the United Nations Disaster Relief Co-ordinator (UNDRO)

UNDRO commissioned this manual to present a broad strategy for disaster mitigation. It is a practical guide that makes scientific and technical information on natural disasters and mitigation strategies available to government policy-makers and planners. In this manual, disaster-resistant building is considered in the context of a much wider process of mitigating disasters. The chapters of this manual are: objectives and policy framework; government administration for risk reduction; risk assessment; planning and decision-making; effective implementation of a strategy for risk reduction; conclusions; hazard assessment; hydrological aspects; geological aspects; vulnerability analysis; options for risk reduction; risk reduction measures for areas affected by hydrological and geological hazards; and cost of risk-reduction measures.

UNDRO; 1991; 164pp; $30
Available from: UNDRO, Palais des Nations, 1211 Geneva 10, Switzerland.

Planning for Human Settlements in Disaster-prone Areas
UNCHS(HABITAT)

The physical vulnerability of human settlements to natural disasters is the focus of this report and it is intended to raise the issue of disaster mitigation for officials engaged in settlement planning. Characteristics of different natural hazards and their impact on settlements are discussed. A basic framework for disaster mitigation planning for human settlements is provided. The publication is only concerned with physical vulnerability, and does not consider the wider relationship between vulnerability and social and economic factors.

UNCHS(HABITAT); 1983; 29pp; free
Available from: UNCHS(HABITAT), PO Box 30030, Nairobi, Kenya.
Also available in French and Spanish.

Post-disaster shelter

Shelter After Disasters: guidelines for assistance
Office of the United Nations Disaster Relief Co-ordinator (UNDRO)

Despite the fact that this is now more than 10-years-old, this manual still provides comprehensive policy and programme guidelines on post-disaster shelter provision. It covers disaster preparedness, emergency relief, reconstruction and preventive measures. These guidelines are intended for officials and technical staff responsible for planning and executing post-disaster shelter programmes. They are not intended for use at a local level of implementation, since it is argued that local guidelines need to be formulated in the context of local conditions. One feature of these guidelines is the emphasis on assessing shelter needs from the perspective of the survivors. The two main sections of this report are emergency shelter and post-disaster housing. The former covers the following main issues: the needs and resources of survivors; allocation of roles to assisting groups; the assessment of survivors' needs; the role of emergency shelter; shelter strategies; and contingency planning (preparedness). The second section on post-disaster housing covers: reconstruction; relocation of settlements; land tenure and land use; and housing finance. The main conclusions of the report are summarized in 14 key principles. The appendices include summaries of the case studies used in the report, and information on relevant books, periodicals, films and organizations (although this information is now 10-years-old).

UNDRO; 1982; 82pp; $20
Available from: UNDRO, Palais des Nations, 1211 Geneva 10, Switzerland.
Also available in Spanish.

Housing and Culture after Earthquakes
Aysan, Y. and Oliver, P.

This guide was prepared from the results of a research project carried out from 1982-6 on the cultural aspects of housing following the Gediz earthquake of 1970 in western Turkey. The guide focuses on the socio-cultural acceptability of post-disaster housing policies in the following areas: emergency shelter; relocation; permanent housing; house design; financing housing; reducing vulnerability; risk perception; access to old villages; housing families; and modernization through reconstruction. The implementation of each of these policies in Gediz is described. The local implications of these policies in Gediz are then discussed, and the social and cultural problems resulting from such policies are identified. Policy recommendations are provided on how such problems could be resolved in future. This is probably the first report to consider in depth the social and cultural problems resulting from post-disaster housing provision. It combines general insights with detailed case study material and is still the only book that has attempted to link cultural concerns to low-cost buildings as part of reconstruction planning.

Oxford Polytechnic; 1987; 70pp
Available from: School of Architecture, Oxford Brookes University, Gipsy Lane, Oxford OX3 0BP, UK.

Disasters and the Small Dwelling: perspectives for the UN IDNDR
Aysan Y. and Davis, I. (eds)

The proceedings of the conference on Disasters and the Small Dwelling, held in Oxford in 1990, form the basis of this book. This conference served as an opportunity to review the progress, successes and failures of post-disaster housing, and to provide recommendations for further development. The book contains 26 papers which cover recent experiences of post-disaster shelter and housing provision, drawn from a wide variety of disaster situations. It also contains overviews which review the current situation, set against progress during the past decade.

James and James; 1992; 264pp; £35/$60
Available from: James and James Ltd, 5 Castle Rd, London NW1 8PR, UK.

Transition Housing For Victims of Disasters
Office of Foreign Disaster Assistance (OFDA)

The immediate shelter needs of disaster victims are addressed in this manual. The focus of the book is on programme planning for contractor-built post-disaster housing programmes. A key concern is with the progression from transitional housing to permanent housing. The manual is organized in three parts. Chapters I and II review the character of the project situation following disasters. Chapters III to VIII review the technical issues set by different natural hazards, including floods. Chapters IX and X are concerned with project preparation and implementation.

OFDA; 1981; 245pp;
Available from: OFDA, USAID, Washington DC, 20523, USA.

2. Training and communication

This section contains training manuals on communication methods for development with a particular focus on the community level. Many of these come from the field of health care or agricultural extension. Nevertheless, they are directly relevant to the problem of communicating safe building techniques because they address the issue of how to communicate technical material in a straightforward, culturally appropriate manner.

Communication Skills for Rural Development
MacDonald, I. and Hearle, D.

The purpose of this book is to provide practical information on communication skills for extension workers in rural communities. The text is written in a very accessible manner and is accompanied by useful diagrams and exercises. The manual covers communication techniques, running training courses and the use of audio-visual materials. Considerable attention is also given to possible problems arising from the social position of an extension worker in rural communities, the planning and evaluation of programmes and rural development strategies. The need for community participation in all stages of rural extension programmes is stressed by the authors.

Evans; 1984; 128pp; £5.50
Available from: Intermediate Technology Publications, 103/105 Southampton Row, London WC1B 4HH, UK.

Guide to Extension Training
Oakley, P. and Garforth, C.

The basic principles of extension work are set out and a key concern of the authors is to establish dialogue between extension workers and rural people. They stress that extension work has to be seen as part of the wider development process and that the social and cultural background of rural societies must be properly appreciated by extension workers. The manual provides guidelines on communication techniques and how and when to use various types of mass media and audio-visual aids. The value of individual and group methods of extension work are assessed and guidelines provided. Attention is then given to the role of extension agents and the methods and skills required of them. The need for careful planning and evaluation of extension programmes is stressed and guidelines are provided on how to implement programmes. Finally, there is a focus on special target groups for extension work such as women or the landless, and discussion of the problems and issues that arise in extension work with these groups. This is an excellent guide that highlights the extent to which extension training for rural development has developed compared with the rudimentary stage of building improvement programmes, and shows how many common pitfalls can be avoided.

FAO; 1985; 144pp; $19
Available from: Food and Agricultural Organisation, Viale delle Terme di Caracalla, 00100 Rome, Italy.

Teaching and Learning with Visual Aids
Programme for International Training in Health (INTRAH)

Guidelines for using low-cost visual aids are provided in this is practical manual. The book is especially intended for the training of community health workers, health trainers and family planning workers in Africa and the Middle East. However, it will be of value to community and extension workers in general, who need to communicate technical information to a wider audience. The book contains five main sections: when to use visual aids; deciding what kind of visual aid to use; planning and making your own visual aids; production skills; and using visual aids in a training or health education session. The book is written in simple English, and contains hundreds of examples and suggestions for making visual aids. No previous knowledge of using or making visual aids is assumed.

Macmillan/INTRAH/TALC; 1987; 290pp; £4.25
Available from: Teaching Aids at Low Cost (TALC), PO Box 49, St Albans, Hertfordshire AL1 4AX, UK.

Tools for Community Participation: a manual for training trainers in participatory techniques
Srinivasan, L.

Participatory techniques for training are promoted in this comprehensive manual as a means of encouraging community participation in development programmes. Part I of the book covers: community participation in development; planning a participatory training programme; organizing the workshop – resources and logistics; designing the participatory workshop; simple daily evaluation techniques and activities; and follow-up planning – putting participation into daily practice. Part II of the book contains 39 participatory training activities. These exercises are designed to be adapted by trainers to their own specific context. This manual is especially concerned with the participation of women.

PROWESS/UNDP; 1990; 176pp; £15.00
Available from: Intermediate Technology Publications, 103/105 Southampton Row, London WC1B 4HH, UK.

Community Participation – A Trainer's Manual
UNCHS(HABITAT)

This manual provides general guidelines on the design, delivery and evaluation of training programmes. It is intended primarily to assist those involved in the promotion of community participation in low-cost housing programmes to set up training programmes. It also serves as a general trainers' guide to the training modules produced under the UNCHS/DANIDA training programme on community participation (some of these modules have been included elsewhere in this compendium). The manual covers the following main areas: conditions for effective learning; key principles of training programme design; advantages and disadvantages of the main modes of communication in training; evaluating training programmes. It also includes an extensive Toolkit section which contains the following: a summary of different training methods, such as lectures, discussion, exercises; the use of media in training, including the use of video; a large number of training exercises, such as role-play exercises; and checklists for the evaluation of training programmes.

UNCHS(HABITAT); 1988; 149pp; free
Available from: UNCHS(HABITAT).
Also available in Spanish.

Project Support Communication
UNCHS(HABITAT)

The communication needs of low-income housing projects are addressed by these four training modules. The four modules are: 1. basic principles; 2. meetings; 3. documents; and 4. audio-visuals. Each training module is intended to form the basis of a three- to five-day training workshop and can be used independently or as part of a series. The major theme running throughout each module it the importance of participatory communication. The course is targeted at project managers and staff involved in low-income housing projects. Module one covers the basic principles of communication for low-income housing projects and suggests methods of evaluation and monitoring the effectiveness of participatory communication. Module two is concerned with the organization, purpose and effectiveness of meetings and explains the basic principles of interpersonal communication. Module three covers the role of written documents, such as reports and minutes, in the communication process of projects. Module four explains how audio-visual materials can be used for project support communication. It provides useful guidelines on how and when to use selected media, including video.

UNCHS(HABITAT); 1986; 224 pp; free
Available from: UNCHS(HABITAT).
Also available in French and Spanish, except module three.

Técnicas Participativas Para La Educación Popular: tomo 1
ALFORJA

Written in Spanish, this is an easy-to-follow, well-illustrated guide to participatory education techniques. The book contains many exercises for use by group facilitators in order to encourage teamwork, improve communication skills, and analyse and solve problems.

ALFORJA; 183pp; $13
Available from: Centro de Estudios Y Publiciones, ALFORJA, Apartado Postal 369, San José, Costa Rica.

Audio-visual Communication Handbook
Pett, P.

This handbook is designed to assist development workers plan, produce and use audio-visual aids for communication, using local materials. It contains 10 appendices which cover such areas as planning, evaluation and sources of information.

World Neighbors; $4
Available from: World Neighbors, 4127 NW 122 St, Oklahoma City, OK 73120-8869, USA.

Helping Health Workers Learn
Werner, D. and Bower, B.

Written by the authors of *Where there is no Doctor*, this book is directed at community-based health workers. Simple English is used together with hundreds of drawings, diagrams and photographs. The focus of the book is on training health workers on methods and techniques of teaching health care. A central concern of the authors is that health workers must be aware of the relationship between health care and political empowerment. This manual is a key text on training health workers, providing valuable insights that can usefully be applied to other community-based projects, including community-based building projects.

The Hesperian Foundation; 1982; 632pp; £5.50
Available from: The Hesperian Foundation, PO Box 1692, Palo Alto, CA 94302, USA, and TALC, UK.
Also available in Spanish and Portuguese.

Teaching Health Care Workers: a practical guide
Abbatt, F. and McMahon, R.

This is a comprehensive guide for those involved in the training of village health workers. It is written in a straightforward, easy-to-follow style with numerous illustrations. It follows a clear methodology of training centred on reaching decisions on what health workers should actually be learning through analysis of their tasks. Guidelines are given on the planning of training programmes, and on the teaching and assessing of communication skills, attitudes, decision-making skills, factual knowledge and manual skills.

Macmillan; 1985; 249pp; £4.25
Available from: TALC, UK.

Education for Health: a manual on health education in primary health care
World Health Organization

This is another comprehensive book on training in primary health care. It is intended to assist community health workers provide appropriate health education. It includes a chapter on communication techniques.

WHO; 1988; 261pp; SwFr 34
Available from: WHO, Avenue Appia, CH 1211, Geneva 27, Switzerland.

Teaching Adults
Rogers, A.

Although not specifically concerned with education for development, this book presents an approach to teaching that has wide applicability. The author presents a participatory approach to teaching in which the teacher draws heavily on the experience and knowledge of the learners. Although this approach is well-established the book does provide many practical ideas for teaching.

Open University Press; 1986; 197pp;
Available from: Open University Press, Milton Keynes, UK and Sterling Press, L-10 Green Park Extension, New Dehli 110.016, India.

3. Disaster-resistant construction

This section contains technical documents that are primarily concerned with analysing the resistance of small buildings to earthquakes or strong winds. None of the documents selected are specialized engineering studies. Most can be understood by people without an engineering or technical background.

Earthquakes and general

Small Buildings in Earthquake Areas
Daldy, A.F. for the Building Research Establishment

Probably the first study of this topic, it was written for builders and others engaged in the actual construction of small buildings in earthquake areas, rather than for architects and engineers involved in building design. This is not a training manual but more of a simplified technical text book on earthquake-resistant building principles. No consideration is made of adapting the general principles to local housing forms and conditions. Minimal attention is given to the social and economic consequences of technical changes recommended to ensure greater safety. The book explains the basic principles of earthquake-resistant design and then explains the specific building principles for the following: foundations; walls of earth; walls of block, bricks or stone; buildings with concrete walls; buildings with squared roofs; floors; chimneys and flue-pipes; termites; and construction materials.

BRE; 1972; 41pp; £2.00
Available from: Building Research Establishment, Garston, Watford, Hertfordshire WD2 7JR, UK.

Protection of Educational Buildings Against Earthquakes: a manual for designers and builders
Arya, A.S.

This manual is concerned with the construction of earthquake-resistant, non-engineered educational buildings, although the author suggests that the principles recommended in the manual can also easily be modified for normal housing. The protective measures suggested are simple and low-cost, and based on the use of locally available materials. The manual stresses how local materials, normally considered unsuitable for earthquake-resistant building, can be made more resistant if certain techniques are followed. A large number of diagrams are used to illustrate protection measures. Sections covered in the manual include: earthquakes; earthquake impact; seismic conditions for building design; site considerations; general principles regarding building form; materials and quality of construction; wood buildings; masonry buildings; low-quality masonry and adobe construction; and strengthening existing buildings.

UNESCO; 1987; 67pp; $2.00
Available from: UNESCO, Box 967, Prakanong Post Office, Bangkok 10110, Thailand.

Earthquake Protection
Coburn, A.W. and Spence, R.J.S.

Methods for protecting settlements, people and buildings against the effects of earthquakes are reviewed in this book. It is intended for all those concerned with earthquake protection and deals with a range of technical and managerial matters. It is written in a style that should be accessible to those without an engineering background. The book covers the management of earthquake emergencies, recovery and reconstruction, assessment of earthquake hazards, techniques and standards of construction, economic planning measures, and social education for earthquake risk. The book draws on projects of earthquake protection in Turkey, Yemen, Italy, Japan, USA and Mexico, and on earthquake vulnerability studies carried out by the authors.

John Wiley; 1992; 332pp; £34.95/$74.50
Available from: John Wiley and Sons Ltd, Baffins Lane, Chichester, West Sussex, PO19 1UD, UK.

Simplified Building Design for Wind and Earthquake Forces
Ambrose, J. and Vergun, D.

Although it focuses on building in the USA rather than low-cost building in developing countries, this book provides a valuable general textbook for building designers. It is intended primarily for people involved in building design who do not have extensive training in engineering and structural theory. The book contains the following chapters: wind effects on buildings; earthquake effects on buildings; lateral load resistance of buildings; lateral load-resisting systems; design for wind and earthquake effects; wind and earthquake effects on foundations; references. There is also a section on study aids, intended for people using the book as a self-study textbook.

John Wiley; 1990; 307pp; £43.95/$66.50
Available from: John Wiley and Sons Ltd, Baffins Lane, Chichester, West Sussex, PO19 1UD, UK.

General Guidelines for Planning, Designing and Construction Earthquake-resistant School Buildings
Gupta, S.P.

This booklet is a synthesis of available published literature on earthquake-resistant school building construction including the author's own recommendations after observation of several school buildings destroyed in earthquakes. It has been especially prepared to cater for the needs of Asia-Pacific countries. Earthquakes and their effects are discussed and the general planning, site considerations and suitability of materials commonly used (bricks, random stone, hollow block masonry and timber) are covered. Many school buildings in developing countries particularly in rural areas are constructed more or less in a traditional manner and generally not designed to be earthquake-resistant. It is these kind of school buildings which need more attention and on which this book focuses.

ADPC-AIT; 1992; 34pp
Available from: Asian Disaster Preparedness Centre, Asian Institute of Technology, PO Box 2754, Bangkok 10501, Thailand.

Building in Earthquake Areas
Building Research Establishment

This is a highly technical report that is intended as an introduction to the design and construction of earthquake-resistant buildings. It is intended for engineers and architects.

BRE; 1972; 21pp; £2.00
Available from: Building Research Establishment, Garston, Watford, Hertfordshire WD2 7JR, UK.

School Buildings and Natural Disasters
Vickery, D.J.

The risk of school buildings to natural disasters are analysed in this report and policy recommendations for mitigating this risk are provided. It highlights the need for safer construction although does not go into any detail on disaster-resistant building techniques.

UNESCO; 1982; 85pp; $2
Available from: UNESCO, 7 Place de Fontenoy, 75700 Paris, France.

Edificios Escolares y Desastres Naturales Estudio de Caso Sobre Mexico y la Zona Centroamericana
Ruiz Gomez, S.E.

This report is concerned with protecting school buildings in Central America from earthquakes. It provides detailed guidelines on simple earthquake-resistant construction techniques. Although intended primarily for school buildings, the techniques recommended are appropriate to other forms of non-engineered structures.

UNESCO; 1987; 143pp; $2
Available from: UNESCO, 7 Place de Fontenoy, 75700 Paris, France.

Strong winds

Cyclone-resistant Housing for Developing Countries
Building Research Establishment

General guidelines for the location and structural design of low-cost housing to resist cyclones are provided in this publication. In addition, specific structural details are given which emphasize the importance of strong connections between components for wind-resistant housing. The publication brings together the results of studies made by the BRE on the effects of wind on buildings in various developing countries and attempts to establish general principles that can be applied throughout the world. This is a purely technical document in which no consideration is given to any other relevant factors involved in the construction of cyclone-resistant housing. However, the general principles provide important background information.

BRE; 1988; 33pp; £8.00

Available from: Building Research Establishment, Garston, Watford, Hertfordshire WD2 7JR, UK and Intermediate Technology Publications, 103/105 Southampton Rd, London, WC1B 4HH, UK.

Buildings and Tropical Windstorms. Overseas Building Notes 188
Building Research Establishment

This is a short technical document concerned with building design in areas affected by tropical windstorms. It is intended for architects and engineers and some knowledge of the basic procedures of structural design is assumed. The general characteristics of tropical storms and the damage they cause to buildings are briefly described. Methods of calculating both the wind loads on buildings and the internal pressures in buildings are suggested. A comprehensive bibliography for further reference is provided.

BRE; 1981; 17pp; £2.00

Available from: Building Research Establishment, Garston, Watford, Hertfordshire WD2 7JR, UK.

Typhoon-resistant School Buildings for Vietnam
Macks, K.J.

This is a detailed technical analysis of the impact of typhoons on school buildings in Vietnam. It both assesses the technical causes of typhoon damage and provides recommendations for typhoon-resistant buildings. General principles for improved buildings are explained through technical information and diagrams. Although the focus of the report is on school buildings, the general principles will also be applicable for typhoon-resistant housing.

UNESCO; 1987; 110pp; $2

Available from: UNESCO, PO Box 967, Prakanong Post Office, Bangkok 10110, Thailand.

Cyclone Structural Testing Station – Technical Reports
James Cook University

The Cyclone Structural Testing Station has produced more than 30 technical reports on the effects of cyclones on small buildings in the Pacific region and on the testing of houses. These include: *No. 6. Recommendations for the testing of roofs and walls to resist high wind forces; No.7. Connections and fastenings for domestic construction; and No. 29. A discussion of criteria for the structural design of buildings to resist tropical cyclones.*

James Cook University; from 1978; A$20 each

Available from: Cyclone Structural Testing Station, James Cook University, Townsville QLD 4811, Australia.

4. Manuals and guidelines on construction

The following are only a selection of the numerous training manuals on simple techniques for safe building that have been produced following disasters. Most of these manuals are difficult to obtain as time elapses following the disaster and for this reason have not been included in the compendium. In any case the value of many of these manuals is that they were directed at specific communities with specific housing needs and traditions and may be less useful, and possibly harmful, if applied to different contexts those originally intended. Some examples of training manuals can be ordered directly from INTERTECT (see INTERTECT's publications list). INTERTECT have helped in the production of several simple manuals, especially on earthquake-resistant building in Latin America such as SENA's *Manual Para Instructores* which is listed below as an example.

Earthquakes and general

Disaster-resistant Construction for Traditional Bush Houses: a handbook of guidelines.
Boyle, C.

This manual is concerned with the type of traditional housing commonly found on certain islands in the South Pacific. Safer building techniques are explained in a straightforward manner and useful sketches accompany the text. No new building techniques are introduced. Rather, it is recommended that traditional building techniques should be more rigorously adopted, since if followed correctly they can provide adequate protection against natural hazards. The problems caused by strong winds, earthquakes and floods are each considered in turn, and general principles provided on hazard-resistant construction.

Australian Overseas Disaster Response Organization; 1988; 70pp;
Available from: Australian Overseas Disaster Response Organization, PO Box K425, Haymarket 2000, Australia.

Construcciones Sismo-resistentes: manual para instructores
SENA/INTERTECT

This practical manual for building instructors explains earthquake-resistant building techniques. These techniques are explained in a simplified form and a large number of drawings are used. There are 10 main sections to the manual: 1. qué son los terremotos; 2. como afecta el terremoto a una casa; 3. principios básicos de construcción sismoresistente; 4. como construir una casa segura de ladrillo; 5. daños típicos; 6. como inspeccionar y evaluar una casa dañada por el sismo; 7. formulario para evaluación de daños; 8. como reparar una casa de ladrillo; 9. como reparar una casa de adobe; and 10. como reparar daños típicos.

SENA; 1983; 60pp; $20

Available from: SENA, Apartado Aereo 1280, Popayon, Colombia and INTERTECT, 3511 North Hall St, Suite 302, Dallas, Texas, USA.

Construcciones Sismo-resistentes: manual técnico de capacitación

SENA

More technical information is provided in this SENA manual than *Construcciones Sismo-resistentes: manual para instructores*. It contains the following sections: Part 1. Generalidades – causas y effectos; principios basicos de la sismo-resistencia; Part 2. Caracteristics tecnicas – preliminares; cimentacion y desagues; muros; elementos estructurales des confinamiento; cubiertas; piôs, instalaciones y acabados.

SENA; 1984; 119pp;

Available from: SENA, Apartado Aereo 1280, Popayon, Colombia.

Como Arreglar Nuestra Casa et al
United Nations and Junta Nacional de la Vivienda

This series of booklets was produced following the Equadorian earthquake of 1987.

JNV/UNCHS; 1987-90; 30pp each
Available from: JNV, Avienida 10 de Agosto, 2270 y Cordero, PO Box 3244, Quito, Ecuador.

Strong winds

Jack Hammer Series
INTERTECT

The *Jack Hammer Series* contains the following publications, each of which is a simple training manual using cartoons for building to resist strong winds: 1. *Will your house stand up?* 2. *How to make a safe wooden house.* 3. *How to make a safe concrete house.* 4. *How to make a safe block and steel house.* 5. *Improving a nog house.* 6. *Improving a wooden house.* 7. *Improving a block and steel house.*

INTERTECT; 1983; $4.50 each
Available from: INTERTECT, 3511 North Hall St, Suite 302, Dallas, Texas, USA.

Hurricanes and Houses: safety tips for building a board house.
Construction Resource and Development Centre, Jamaica

CRDC produced this manual to provide guidelines on building low-cost hurricane-resistant houses in the Caribbean. The basic principles of safe construction are explained through straightforward messages and clear illustrations.

CRDC; 1988; 18pp;
Available from: CRDC, 11 Lady Musgrave Ave, Kingston, Jamaica.

When You Build a House – a manual of construction details for Caribbean houses
Robinson, E.H.

This is a short, simple manual on house design and construction with an emphasis on protection from strong winds. Clear drawings are provided, along with brief explanations. These show the basic principles for construction to resist strong winds.

Caribbean Conference of Churches; 13pp; US $4
Available from: Caribbean Conference of Churches, PO Box 616, Bridgetown, Barbados, West Indies.

5. Building materials

This section contains key texts on the production, building techniques and qualities of various types of building materials. The focus is on low-cost building materials that can be produced using local materials. Most books are of a general nature covering a wide range of building materials, although some specialist handbooks on earth construction have also been included given the importance of earth as a low-cost building material.

General handbooks

Appropriate Building Materials: a catalogue of potential solutions
Stulz, R. and Mukerji, K.

This is a very comprehensive sourcebook on building materials for low-cost construction. It provides technical information on all the main types of building materials and practical examples of their use in construction. It is well-illustrated with photographs and drawings. The book is intended primarily for building practitioners, but will be valuable to anyone engaged in low-cost construction such as architects, engineers, educational institutions, building material producers and project managers. The main sections of the book contain fundamental information on building materials – 19 different materials are covered; and building elements – foundations, floors and ceilings, walls, roofs, and building systems; protective measures. Examples are provided of foundation materials; floor materials; wall materials; roof materials; building systems – this includes a section on earthquake-resistant mud/bamboo structures and a section on timber houses for flood areas; annex – this includes a section on machines and equipment, a list of 147 organizations involved in appropriate building materials around the world with a brief explanation of their expertise, and a substantial bibliography. An index is also provided for quick reference.

Intermediate Technology Publications/ SKAT/GATE; 1988; 430pp; £12.50
Available from: Intermediate Technology Publications, 103-105 Southampton Row, London WC1B 4HH, UK.

A Compendium of Information on Selected Low-cost Building Materials
UNCHS(HABITAT)

The compendium is concerned with promoting building materials that can be made from locally available raw materials and with relatively simple low-cost production methods and serves as a very useful source book on the technologies for the production of building materials for use in low-cost housing. The compendium concentrates on five different categories of materials: burnt-clay bricks and tiles; soil construction; low-cost binders; fibre concrete roofing; and timber. For each of these materials, detailed information is provided on raw materials, production technologies, performance standards and recent innovations, and brief technical details are also given on the use of these materials in building. The compendium also lists those organizations that are involved in manufacturing and supplying equipment for small-scale building production and those involved in technology transfer. A bibliography is provided, together with suggestions for further reading.

UNCHS(HABITAT); 1988; 106pp; free
Available from: UNCHS(HABITAT), PO Box 30030, Nairobi, Kenya.

Building materials production

Small-scale Building Materials Production in the Context of the Informal Economy
UNCHS(HABITAT)

The problem of the shortage of low-cost building materials in developing countries is addressed in this report. It suggests ways in which locally organized, small-scale building materials production using locally available raw materials can improve this situation.

UNCHS(HABITAT); 1984; 40pp; free
Available from: UNCHS(HABITAT), PO Box 30030, Nairobi, Kenya.

The Use of Selected Indigenous Building Materials With Potential for Wide Application in Developing Countries
UNCHS(HABITAT)

This report examines the factors which act as constraints to the production and use of indigenous building materials and identifies measures which can be undertaken to overcome them. The report draws on 10 case studies and focuses especially on the following materials: lime; natural, lime and artificial pozzolanas; blended cements; gypsum-based binders.

UNCHS(HABITAT); 1985; 70pp; free
Available from: UNCHS(HABITAT), PO Box 30030, Nairobi, Kenya.

Bibliography on Small-scale Building Materials Production
UNCHS(HABITAT)

This bibliography is intended as a complement to other UNCHS(HABITAT) reports on promoting small-scale building materials production. It focuses on the following materials: 1. fire-clay bricks and tiles; 2. cement and concrete products; 3. low-cost binders; 4. soil; 5. timber; 6. fibre-concrete roofing. The bibliography includes an annotated section on several key texts on building materials production.

UNCHS(HABITAT); 1989; 122pp; free
Available from: UNCHS(HABITAT), PO Box 30030, Nairobi, Kenya.

Earth construction

Earth Construction Technology four Vols.
UNCHS(HABITAT)

These four manuals are intended for professionals dealing with projects on earth construction. The four manuals are: 1. *Basic Principles of Earth Application*; 2. *Design and Construction Techniques*; 3. *Production of Rammed Earth, Adobe and Compressed Soil Blocks*; and 4. *Surface Protection*. All these manuals are intended to be basic reference materials from which builders' training manuals can be produced.

UNCHS(HABITAT); 1986; 206pp in total; free
Available from: UNCHS(HABITAT), PO Box 30030, Nairobi, Kenya.

Building With Earth: a handbook
Norton, J.

Both technical information on the performance of earth and practical guidelines on building techniques are covered in this manual on the use of earth as a building material. It should be recognized that these guidelines do not pay any specific attention to resisting earthquakes or strong winds. The main sections of the handbook are: general characteristics; soil types; soil analysis; performance and performance testing; wall construction; roofs, foundations and floors; stabilization; renders; other considerations; and select bibliography.

IT Publications; 1986; 68pp; £6.50
Available from: Intermediate Technology Publications, 103-105 Southampton Row, London WC1B 4HH, UK.

Construire en Terre
CRATerre

CRATerre has produced this detailed practical guide on the characteristics of earth as a building material and on how to use it in construction. The chapters of the book are: introduction; le pisé; façonnage direct et bauge; l'adobe; briques de terre comressées; analyse des sols; caractéristiques du matériau terre; stablization; techniques mixtes; tortures en terre; enduits et peintures; les malaxeurs; du nouveau au CRATerre; and construire en terre au pérou. There is also a list of organizations with expertise in earth construction and a bibliography.

CRATerre; 1985; 287pp; FF105
Available from: CRATerre, Centre Simone Signoret, BP 53F-38090, Ville Fontaine, France.

Bibliography on Soil Construction
UNCHS(HABITAT)

This is a comprehensive bibliography on soil construction, containing about 6000 entries. The bibliography contains a general section, a section divided into specific subject areas in relation to soil construction technology, and a section of selected annotated entries.

UNCHS(HABITAT); 1989; 173pp; free
Available from: UNCHS(HABITAT), PO Box 30030, Nairobi, Kenya.

6. Services for human settlements

This section includes key texts on the provision of services for housing. Further information on books with a specific focus on water and sanitation can be found in Intermediate Technology's *Books by Post* catalogue, which contains an extensive list of relevant publications.

Services for Shelter: infrastructure for urban low-cost housing
Cotton, A. and Franceys, R.

Services for Shelter is a comprehensive manual that presents and analyses a strategy for the planning, implementation, operation and maintenance of services for low-cost housing. It is a practical guide rather than a highly technical engineering manual. A major concern of the manual is on community participation in the process of services provision. The book contains the following chapters: people, shelter and services; ground preparation; drainage; access and circulation layout; water supply; sanitation; solid waste management; power supply; interaction; involvement and implementation. There is also a substantial references section on services for low-cost settlements.

Liverpool University Press/Fairstead Press; 1991; 147pp; £17.50
Available from: Liverpool University Press, PO Box 147, Liverpool L69 3BX, UK.

Services for the Urban Poor: a select bibliography
Franceys, R. and Cotton, A.

This bibliography covers the following: infrastructure planning; site preparation; drainage; roads and access; water supply; sanitation; and solid waste management.

Intermediate Technology Publications; 1993; 84 pp; £15
Available from: Intermediate Technology Publications, 103-105 Southampton Row, London WC1B 4HH, UK.

UNCHS(HABITAT) Training Modules

The following documents are all UNCHS(HABITAT) training modules on community participation and the provision of or improvement of services for housing. Each of these training modules provide a framework for a training workshop, appropriate technical information and suggested training exercises. These are intended for 10 to 20 participants and to last two to four days. Each module also contains guidelines for trainers and a bibliography.

Water Supply in Low-income Housing Projects: the scope for community participation
(1989; 60pp)

Sites and Services Schemes: the scope for community participation
(1984; 44pp; also available in French and Spanish)

Community Participation in Low-cost Sanitation
(1986; 81pp; also available in French and Spanish)

Community Participation in Squatter Settlement Upgrading
(1985; 44pp; also available in French and Spanish)

Community Participation and Low-cost Drainage
(1986; 59pp; also available in French and Spanish)

All the above modules are available from: UNCHS(HABITAT), PO Box 30030, Nairobi, Kenya.

7. Project planning and evaluation

This section contains material relating to methods of planning and evaluating development programmes. Most of the entries are not specifically concerned with housing programmes but they all provide general principles that can be used for the design of projects and for evaluation work in the context of building improvement programmes.

General project planning

Guidelines for the Preparation of Shelter Programmes
UNCHS(HABITAT)

This UNCHS(HABITAT) document provides a detailed and comprehensive set of guidelines to assist agencies in the provision of low-cost shelter. It is aimed primarily at officials from national and local governments. It provides a step-by-step guide for all stages of shelter programmes: from needs analysis to objectives; assessment of resources; affordability, costs and options; matching objectives with resources; formulation of programme; implementation procedures; and monitoring and evaluation. This is useful document for those engaged in large-scale reconstruction programmes.

UNCHS(HABITAT); 1984; 84pp; free
Available from: UNCHS(HABITAT), PO Box 30030, Nairobi, Kenya.
Also available in French and Spanish.

The Field Directors Handbook
Oxfam

This has become established as one of the most comprehensive manuals on development work, providing invaluable guidelines for fieldworkers and decision-makers alike. It offers both practical advice and discussion of the key issues involved on the planning, implementation and evaluation of development programmes. The main areas covered are priority groups, field methodologies, social and economic development, agriculture, health and disaster relief. The chapter on disaster relief includes a section with guidelines on emergency shelter provision.

Oxfam; 1985; 512pp; £12.95
Available from: Oxfam, 274 Banbury Rd, Oxford OX2 7DZ, UK and Intermediate Technology Publications, 103/105 Southampton Row, London, WC1B 4HH, UK.

Social Survey Methods: a guide for development workers
Nicholas, P.

This practical manual provides guidelines on simple, low-cost techniques for gathering social and economic information. No specialist knowledge of research methods is needed to use this guide. The subjects covered in this guide are: surveys and other social research methods; study design – planning and budgeting for the survey; the fieldwork team – recruitment, training and supervision of fieldworkers; form design – methods of collecting and recording information; choosing the sample; techniques for data analysis; and presenting the findings.

Oxfam; 1991; 176pp; £6.95
Available from: Oxfam, 274 Banbury Rd, Oxford OX2 7DZ, UK and Intermediate Technology Publications, 103/105 Southampton Row, London, WC1B 4HH, UK.

Choosing Research Methods: data collection for development workers
Pratt, B. and Loizos, P.

A companion volume to *Social Survey Methods*, this manual is intended to assist development workers with the collection of a wide range of data needed for development work. It considers the broader social issues behind social research and explains and evaluates the different methods of collection currently in use.

Oxfam; 1992; 120pp; £5.95
Available from: Oxfam, 274 Banbury Rd, Oxford OX2 7DZ, UK.

Indigenous Peoples: a fieldguide for development
Beauclerk, J., Narby, J. and Townsend, J.

This guide is useful for project workers and policy-makers engaged in building improvement programmes that incorporate indigenous groups. The guide discusses the main issues that need to be considered in any form of development programme involving indigenous groups.

Oxfam; 1988; 144pp; £5.95
Available from: Oxfam, 274 Banbury Rd, Oxford OX2 7DZ, UK and Intermediate Technology Publications, 103/105 Southampton Row, London, WC1B 4HH, UK.

Evaluation

Partners in Evaluation: evaluating development and community programmes with participants
Feuerstein, M.

Techniques for the evaluation of community programmes in developing countries are carefully explained in this practical manual. It is written in a straightforward style and contains many helpful illustrations. Different techniques of evaluation are introduced and detailed guidelines are given of how to implement them in practice. A major concern of the book is the training of local people to become involved in the evaluation of programmes taking place in their communities. The book will be very useful to anyone wanting to undertake low-cost evaluation of community development programmes, such as in health, agriculture and housing. No prior knowledge of evaluation methodologies is assumed.

Macmillan/TALC; 1987; 208pp; £4.50
Available from: Intermediate Technology Publications, 103/105 Southampton Row, London, WC1B 4HH, UK and Teaching Aids at Low Cost (TALC), PO Box 49, St Albans, Hertfordshire AL1 4AX, UK.

Shelter Upgrading For The Urban Poor: evaluation of Third World experience
Skinner, Taylor and Wegelin (eds)

The evaluation of shelter upgrading programme is explored in this book through 10 detailed case studies from Africa, Asia and Latin America. These are focused particularly on the impact of upgrading on the following: home improvements; land tenure and value; incomes and expenditure patterns; household mobility; community participation and self-reliance; institutional and bureaucratic change; and implications for urban housing policy and development. These case studies refect a wide range of quantitative and qualitative evaluation approaches. Athough not a practical manual, this book would be of value to policy-makers and those planning to undertake evaluation work. It raises crucial issues that need consideration in conducting such work.

Island Publishing House; 1987; 261pp; free
Available from: UNCHS(HABITAT), PO Box 30030, Nairobi, Kenya.

Questioning Practice: NGOs and evaluation
Porter, D. and Clark, K.

This short book serves as a useful introduction to the evaluation of development projects. It identifies and discusses the main issues relating to evaluation, focusing on the NGO programmes in the Pacific region. It is not a practical manual of techniques for carrying out evaluation, but it suggests general principles and guidelines that should be followed in conducting such work.

NZCTD; 1985; 52pp; £4.95
Available from: Intermediate Technology Publications, 103/105 Southampton Row, London, WC1B 4HH, UK.

Listen to the People: participant-observer evaluation of development projects
Salmen, L. F.

The use of participant observation for the evaluation of development programme is promoted in this book. This approach, which is particularly suitable for qualitative evaluation rests on the evaluator's ability to listen to local views and assess their significance. This contrasts with approaches to evaluation based on preformulated questionnaires or sophisticated statistical analysis. This book is concerned with explaining to project workers the advantages and problems with this method and its uses in the design and management of development programmes.

Oxford University Press/World Bank; 1987; 149pp;
Available from: The World Bank, 1818 H St, NW, Washington, DC 20433, USA.

Self-evaluation Manual
World Neighbors

This low-cost manual on evaluation techniques is intended for use in a wide range of community development projects. It is written in a straightforward style and provides practical suggestions illustrated by examples from the field.

World Neighbors; $5
Available from: World Neighbors, 4127 NW 122 St, Oklahoma City, OK 73120-8869, USA.

Evaluating Social Development Projects
Marsden, D. and Oakley, P. (eds)

This collection of discussion papers addresses the problem of how to evaluate the social development dimension of projects, such as community participation, empowerment or local sustainability. It focuses on four main themes: qualitative indicators for evaluation; methodologies for social development evaluation; partnership in evaluation; the role and position of the evaluator.

Oxfam; 1990; 144pp; £6.95
Available from: Oxfam, 274 Banbury Rd, Oxford OX2 7DZ, UK.

Intermediary NGOs: the supporting link in grassroots development
Carroll, T. F.

Intermediary NGOs play a key role, supporting grassroots groups on a local level, and linking them with donors on the international level. Many reconstruction programmes are organized through intermediary NGOs. This book, based on an examination of more than 30 Latin American NGOs, provides criteria and guidelines for the evaluating the performance of intermediary NGOs, and recommends how international donor NGOs might support and interact with these NGOs.

Kumarian Press; 1992; 274pp; $24.95
Available from: Kumarian Press, 630 Oakwood Avenue, Suite 119, West Hartford, Connecticut 06110-1529, USA.

8. Community organization

The material in this section is concerned with community organization, especially the issue of community participation in development projects. Many of the entries have been produced by UNCHS(HABITAT)) which, in recent years, has had a special concern with community participation housing programmes.

Projects With People: the practice of participation in rural development
Oakley, P. et al

This is a major review of the experiences that agencies have had with community participation. It identifies the key elements and issues involved in promoting community participation in development programmes. It contains a number of case studies from a wide range of development projects, and examples of both government and NGO efforts to encourage community participation. The book is recommended for those wanting a comprehensive assessment of the issues surrounding the incorporation of community participation into development programmes.

ILO; 1991; 284pp; £14.50
Available from: Intermediate Technology Publications, 103/105 Southampton Row, London, WC1B 4HH, UK or International Labour Organization, 4 Chemin des Morillons, CH 1211, Geneva 22, Switzerland.

Building Community
Habitat International Coalition

Twenty case studies from around the world form the core of this book on community-based initiatives in the provision of low-cost housing. Each study focuses on a single housing project, identifying and analysing its problems and successes. The introduction to the book focuses on the issues involved in community approaches to housing provision and the conclusion draws together the main lessons arising from the case studies and makes recommendations for future action. The analysis is focused on key linkages between physical provision of housing and its link with communities.

Building Community Books; 1988; 190pp; £12.50
Available from: Intermediate Technology Publications, 103/105 Southampton Row, London, WC1B 4HH, UK.

Initiatives in Low-cost Housing – A Resource Manual
Association of Development Agencies (ADA) and Council of Voluntary Social Services (CVSS)

This manual is concerned with community-orientated initiatives in low-cost housing. Housing provision is seen in the context of broader social development, rather than in purely material terms. It is a source of ideas and information, drawn from previous community housing initiatives in Jamaica. The manual is especially concerned with the informal sector, and covers such issues as strengthening community organizations, lobbying and co-operative efforts. This manual will be particularly useful for NGOs and community-based organizations engaged in low-cost housing.

ADA/CVSS; 1990; 52pp;
Available from: ADA, 14 South Avenue, Kingston 10, Jamaica.

The Community Builders: a practical guide where people matter
Simey, I. for GATE/GTZ, Germany

The process of building as a community is described step-by-step in this guide. It is based on the author's experience of community-based building projects in Lesotho. The guide is very comprehensive, covering both practical considerations such as cost estimation and financial organization and considers how to involve people in decision-making in all stages of the project.

Vieweg, Germany; 1989; 188pp; £11.95
Available from: German Appropriate Technology Exchange (GATE), Dag-Hammarskjoeld Weg 1, PO Box 5180, D-6236 Eschborn 1, Germany.

The Role of Community Participation in Human Settlements Work
UNCHS(HABITAT)

The basic principles of community participation and ways in which it can contribute to housing programmes are described in this short report. It briefly reviews different types of community participation that can be used in housing programmes and discusses possible constraints on community participation.

UNCHS(HABITAT); 1986; 24pp; free
Available from: UNCHS(HABITAT), PO Box 30030, Nairobi, Kenya.

Community Participation in the Execution of Low-income Housing Projects
UNCHS(HABITAT)

This report serves as a brief, but useful introduction to the role of community participation in low-income housing projects. Both the possibilities and limitations for community participation are discussed, together with the main problems that may be encountered in implementing community participation and means of overcoming them. This is not a practical guide but rather a report that draws attention to issues related to community participation in low-income housing projects.

UNCHS(HABITAT); 1984; 31pp; free
Available from: UNCHS(HABITAT), PO Box 30030, Nairobi, Kenya.
Also available in French and Spanish.

Community Leadership and Self-help: a technical report
UNCHS(HABITAT)

This is an important report which will be of value to anyone involved in community-based housing programmes. It identifies the issue of community leadership as critical to the successful implementation of housing projects. It aims to raise the awareness of the importance of leadership in community development and to encourage external agencies to be sensitive to the way they engage with local leaders and residents. The central concerns of the report are: the nature of community leadership and the questions of their legitimacy and motivation; the leaders' links to the wider socio-political system; the impact of leaders on community development; the relationship between leaders and external agencies; and practical guidelines on working through community leaders.

UNCHS(HABITAT); 1988; 43pp; free
Available from: UNCHS(HABITAT), PO Box 30030, Nairobi, Kenya.

Co-operative Housing: experiences of mutual self-help
UNCHS(HABITAT)

The role of co-operative groups in housing projects are discussed in this publication. General guidelines are given on the organization, development and management of housing co-operative projects, financial management of projects and institutional support. Detailed case studies are provided of four co-operative housing projects from Nicaragua, Ethiopia, Zimbabwe and the Philippines. This book is intended for those involved in the planning and organization of co-operative housing projects.

UNCHS(HABITAT); 1989; 164pp; free
Available from: UNCHS(HABITAT), PO Box 30030, Nairobi, Kenya.

Promoting Organized Self-help Through Co-operative Modes of Participation
UNCHS(HABITAT)

Two major objectives are addressed in this report: 1. to assess the requirements of the co-operative approach to improving human settlements in the light of project experiences. Eight major requirements are put forward and discussed in the light of numerous case studies 2. to identify the training needs in integrated co-operative projects for human settlements in slums and squatter areas and to analyse teaching materials and training programmes which can strengthen participation in co-operative forms of development. It identifies organization and management, self-help building techniques, and finance and legal aspects as problem areas requiring training.

UNCHS(HABITAT); 1984; 61pp; free
Available from: UNCHS(HABITAT), PO Box 30030, Nairobi, Kenya.

Mutual Aid: house construction through building groups – training module
UNCHS(HABITAT)

This training module focuses on the construction of houses by groups of individuals who are assisting each other in building their own homes. It is intended to be used by project staff on shelter projects, such as squatter settlement upgrading projects. The following issues are covered: the advantages of mutual aid housing; organization of mutual aid housing; project design; and finances of mutual aid housing.

UNCHS(HABITAT); 1986; 55pp; free
Available from: UNCHS(HABITAT), PO Box 30030, Nairobi, Kenya.
Also available in French.

Community Participation in Problem-solving and Decision-making. Three Training Modules
UNCHS(HABITAT)

These three modules are: 1. *Basic Principles*. This sets out the basic framework for problem-solving and decision-making, introducing a number of key concepts related to analysing problem situations and preparing action plans. 2. *Leadership*. This focuses on the problems and potentials of leading project staff and community groups in problem-solving and decision-making activities; 3. *Managing Conflict*. This is concerned with coping with conflicts that arise in human settlements projects. Each module includes guidelines for trainers.

UNCHS(HABITAT); 1989; three vols.; free
Available from: UNCHS(HABITAT), PO Box 30030, Nairobi, Kenya.

Catalogue of Training and Information Tools on Community Participation in Human Settlements
UNCHS(HABITAT)

This contains a total of 50 fully annotated entries, most of which are manuals or books but also some audio-visual materials, exercises and newsletters.

UNCHS(HABITAT); 1989; 55pp; free
Available from: UNCHS(HABITAT), PO Box 30030, Nairobi, Kenya.

Co-operative Housing: a bibliography
UNCHS(HABITAT)

Literature on self-help housing groups, as well as formally registered co-operative societies, are included in this bibliography. It is not an exhaustive bibliography, but has been compiled from materials that are readily available. All entries are annotated. The bibliography is divided into four sections: 1. general (16 entries); 2. Africa (33 entries); 3. Asia (13 entries); 4. Latin America (five entries).

UNCHS(HABITAT); 1989; 27pp; free
Available from: UNCHS(HABITAT), PO Box 30030, Nairobi, Kenya.

Bibliography on Community Participation., I and II
UNCHS(HABITAT)

This is a comprehensive, annotated bibliography on community participation in urban and rural housing and in related fields of health and education.

UNCHS(HABITAT); 1983/86; 75/116pp; free
Available from: UNCHS(HABITAT), PO Box 30030, Nairobi, Kenya.

9. Gender issues

The importance of gender issues in planning and running developing programmes are covered in the publications in this section, together with practical guidelines for encouraging greater involvement of women in housing programmes.

Gender and Development: a practical guide
Ostergaard, L. (ed)

The importance of gender awareness in development programmes is discussed in this book. The emphasis is on providing practical methods of involving women more fully in development efforts, rather than an academic discussion of the issues. The volume contains specific section on agriculture; employment; housing; transport; health; and household resource management. The section on housing contains much useful information on the consequences of ignoring gender issues in housing programmes, and discusses the mistaken assumptions in current housing policy.

Routledge; 1992; 220pp
Available from: Bookshops or Routledge, 11 New Felter Lane, London EC4P 4EE, UK and 29 West 35th St, New York, 10001, USA.

Gender Analysis in Development Planning: a case book
Rao, A., Anderson, M.B. and Overhold, C.A. (eds)

The purpose of this training guide is to assist development workers to incorporate gender considerations into development programmes. The book is based around a number of cases which are intended to provide training in developing management strategies and planning and evaluation techniques that are sensitive to gender issues. A set of teaching notes have been produced to accompany the case book.

Kumarian Press; 1991; 102pp/25pp (teaching notes); $18.25/$10.95.
Available from: Kumarian Press, 630 Oakwood Avenue, Suite 119, West Hartford, Connecticut 06110-1529, USA.

Building-related Income Generation for Women – Lessons from Experience
UNCHS(HABITAT)

The specific needs of women in finding training and employment in the construction sector are addressed in this report. It is intended for decision-makers in governments and NGOs who are developing strategies for increasing employment opportunities for women. The report discusses the issues of women's access to training, training processes for women and women's access to employment. While no detailed practical guidelines are given, the report does offer general suggestions for increasing the involvement of women in the construction industry in developing countries.

UNCHS(HABITAT); 1990; 40pp; free
Available from: UNCHS(HABITAT), PO Box 30030, Nairobi, Kenya.

Communication Development and Women's Participation in Human Settlements Management
UNCHS(HABITAT)

This report focuses on the importance of developing effective communication strategies for increasing the involvement of women in urban settlement schemes. It is argued that the particular needs of women are usually overlooked, which can undermine their participation in such schemes. Some general suggestions are given for communicating more effectively to women, although the report is more concerned with discussing issues than providing practical guidelines.

UNCHS(HABITAT); 1988; 64pp; free
Available from: UNCHS(HABITAT), PO Box 30030, Nairobi, Kenya.

Women and Shelter
Joint United Nations Information Committee (JUNIC)

A series of kits have been produced by the JUNIC/NGO Programme Group on Women including this one on shelter issues. The sections of the kit cover the following issues: global human settlement trends; women's access to shelter; women's participation in shelter policies, programmes and projects; women and shelter in emergency situations; women and the construction sector; a resources guide including lists of relevant organizations and a selected bibliography; background papers from various sources. The kit is intended to be used as a foundation for a workshop or discussion groups addressing problems of women and shelter.

JUNIC; 340pp; free
Available from: UNCHS(HABITAT), PO Box 30030, Nairobi, Kenya.
Also available in French and Spanish.

Women in Human Settlements Development and Management
UNCHS(HABITAT)

This is a useful overview of the main types of problems encountered by women in low-cost housing programmes.

UNCHS(HABITAT); 83pp; free
Available from: UNCHS(HABITAT), PO Box 30030, Nairobi, Kenya.

The Role of Women in the Execution of Low-cost Income Housing Projects – Training Module
UNCHS(HABITAT)

The strengthening the position of women in the execution of sites and services schemes and squatter settlement upgrading projects is addressed in this training module. It is intended for project staff involved in such projects. The module covers the following issues: formulating eligibility criterion; recruitment of beneficiaries; settlement planning; planning infrastructure; planning the dwelling; financing housing; women and house construction; self-help building and contractors; project maintenance; and cost recovery.

UNCHS(HABITAT); 1986; 64pp; free
Available from: UNCHS(HABITAT), PO Box 30030, Nairobi, Kenya.
Also available in French and Spanish.

Bibliographic Notes: women and human settlements; the experience
UNCHS(HABITAT)

This contains 38 annotated entries in the following subject areas: general, women in development (12); women, housing policies and housing development (eight); case studies (13); and training (five).

UNCHS(HABITAT); 1989; 7pp; free
Available from: UNCHS(HABITAT), PO Box 30030, Nairobi, Kenya.

10. Financial management

This section contains material relating to all levels of the financial organization of housing programmes, from financing the programme to practical guides on accounting for builders.

Accounting and Book Keeping for the Small Building Contractor
Miles, D.

This manual is intended primarily for managers and owners of small contracting businesses in developing countries. The emphasis is on providing practical ideas and skills for improving the overall management and financial control of such businesses. The guidelines are also relevant for building managers in the public sector. Topics covered include: keeping records; assets and liabilities; basic book keeping; analysis; fixed assests; depreciation; balance sheets; profit and loss accounts; and reading and comparison of accounts. The book ends with a practical exercise on preparing final accounts.

IT Publications; 1979; 190pp; £8.50
Available from: Intermediate Technology Publications, 103/105 Southampton Row, London WC1B 4HH, UK.

Basic Accounting for Small Groups
Cammack, J.

This is a beginners' guide to accounting. It is especially concerned with accounting for small enterprises and organizations in developing countries. It can be used either as a training guide or as a teach-yourself guide.

Oxfam; 1992; 40pp; £3.95
Available from: Oxfam, 274 Banbury Rd, Oxford OX2 7DZ, UK.

Mobilization of Financial Resources for Low-income Groups
UNCHS(HABITAT)

The possibilities of mobilizing the existing financial resources of low-income communities for housing provision are explored in this report. It is intended for those planning and implementing funding systemsb for low-cost systems. It focuses on both the establishment of systems for encouraging and utilizing savings and lending systems. The economic preconditions needed for these systems to operate effectively are also identified. Numerous case studies of systems of channelling savings and loans into housing are provided.

UNCHS(HABITAT); 1989; 31pp; free
Available from: UNCHS(HABITAT), PO Box 30030, Nairobi, Kenya.

Improving Income and Housing: employment generation in low-income settlements
UNCHS(HABITAT)

The inter-relationship between low-income housing and income generation are examined in this publication. It identifies both the potential and constraints for employment in the informal housing sector. This has the benefit of both improving income generation and housing provision. The report contains five chapters. The first serves as a general introduction. Chapter two assesses the potential for employment and income generation in the construction and building-materials production section. Chapter three examines the possibilities for income generation schemes in low-cost housing. Chapter four examines the possibilities for income generation in the provision of services of housing, such as sewage and waste disposal. Chapter five discusses the way in which both governmental agencies and NGOs can support income generation in low-cost housing provision.

UNCHS(HABITAT); 1989; 72pp; free
Available from: UNCHS(HABITAT), PO Box 30030, Nairobi, Kenya.

Community-based Finance Institutions
UNCHS(HABITAT)

This report analyses of the role of community-based finance institutions in mobilizing the resources of low-income groups for the improvement of housing. It focuses on credit unions and housing co-operatives. Credit unions enable low-income groups, which do not have access to conventional institutions, to save and borrow. Housing co-operatives not only mobilize funds but also bring individuals together to act collectively on housing issues. The study is based on three case studies from Jamaica, Kenya and Zambia. The following issues are addressed in the report: use of credit union loans; limitations of short-term loans; linkages to external sources of finance; collaboration with implementing agencies; long-term versus short-term loans; limitations on internally mobilized capital; interest rates; institutional investments; implementation of projects; and serving low-income groups.

UNCHS(HABITAT); 1984; 68pp; free
Available from: UNCHS(HABITAT), PO Box 30030, Nairobi, Kenya.

Credit and Savings for Development: a practical guide
Devereux, S. and Pares, H.

The different types of credit systems for low-income groups are reviewed in this book. Although focused on rural groups, it is also applicable for low-income urban groups. It is intended for people concerned with the planning, implementation and evaluation of credit and savings schemes in developing countries. The first part of the book contains a brief analysis of the rural economy and the situations in which there is a need for credit. It then reviews the financial systems, both formal and informal, that are commonly found in rural areas. The second part of the book examines in detail the issues raised in the implementation of credit and savings programmes. This is under the following headings: access to credit; creditworthiness; default; savings; programme design; participation; and dependency. In the concluding chapter, policy guidelines on credit and savings programmes are recommended.

Oxfam; 1990; 80pp; £4.95
Available from: Oxfam, 274 Banbury Rd, Oxford OX2 7DZ, UK.

How People Can Afford Shelter – Training Module
UNCHS(HABITAT)

The purpose of this training module is to provide guidelines on how to assess what people can actually afford on housing and on effective means of recovering debt. It aims at making community participation effective in determining affordability and organizing cost recovery for low-income housing projects. The module is intended for use by staff of national and local housing authorities and NGOs who are involved in the day-to-day planning and implementation of community-based low-income human settlements.

UNCHS(HABITAT); 1988; 52pp; free
Available from: UNCHS(HABITAT), PO Box 30030, Nairobi, Kenya.

Community Credit Mechanisms – Training Module
UNCHS(HABITAT)

This training module has been prepared as a training course on community participation in the setting up of community credit mechanisms. It provides a framework for the training. The trainer decides what materials to take from this document to give the participants for reading and guidance, and must also adjust the course to local conditions. The following issues are covered: what sort of credit mechanism to use; the establishment of a credit mechanism; the management of a credit mechanisms; and the promotion of a credit mechanism.

UNCHS(HABITAT); 1989; 49pp; free
Available from: UNCHS(HABITAT), PO Box 30030, Nairobi, Kenya.

11. Case studies

A large number of cases studies on post-disaster housing programmes have been written in the past 20 years. Most of these have been written as separate reports or in journals. There is therefore a problem of availability. This section contains publications that are currently available. Some of them are collections of previous case studies and others detailed, published case studies. Some technical case studies are also included. The particular value of case studies lies in their ability to gain a holistic grasp of the difficulties and pitfalls, and successes and failures of programmes. Case study reports also bring together all the elements of projects.

A Study of the State-of-the-art in Earthquake Mitigation Projects: training local builders and the public
Aysan, Y.

This study evaluates programmes and research on improving the earthquake safety of non-engineered buildings in rural areas. In Part 1 projects are examined from Yemen, Guatamala, Colombia and Ecuador. In Part 2, the state-of-the-art in relation to earthquake mitigation projects in the rural areas of Turkey, Iran and Pakistan are discussed. There is a useful annotated bibliography that covers the literature on builder and community training programmes from disaster-prone countries, technical reports on building vulnerability analysis and improvement of non-engineered buildings, and on training methods in rural areas. The appendices contain a number of valuable case studies on builders training programmes and evaluation reports: Fritch, N. and Parker, J. 'A System for Planning Education Materials'; INTERTECT (1983) 'Housing Education in Popayan'; Parker, R. 'The Guatemalan Housing Education Programme'; Mckay, M. 'The Oxfam/ World Neighbors Housing Programme in Guatemala'; Dudley, E. 'Disaster Aid: Equity First'.

Aga Khan Trust for Culture; 1990; 130pp;
Available from: Oxford Centre for Disaster Studies, PO Box 137, Oxford OX4 1BB, UK.

Building for Hope in Costa Rica
Building and Social Housing Foundation

This book contains a collection of selected case studies of low-cost housing programmes in developing countries. It is the proceedings of an international conference on low-cost housing held in Costa Rica in 1991. There are several case studies from the Costa Rica Housing Programme, including one on earthquake-resistant housing. Other cases studies included in the book are: the 1.5 million housing programme of Sri Lanka; the rural housing programme in Malawi; the typhoon-resistant housing programme in the Philippines; urbanization, squatter housing and a civil society organization in Turkey; the housing programme in Chile. The book also includes a directory and description of the organizations which participated in the conference.

Building and Social Housing Foundation; 1992; 166pp; £10
Available from: Building and Social Housing Foundation, Memorial Square, Coalville, Leics LE6 4EU, UK.
This book is jointly written in English and Spanish.

Evaluation of Post-Disaster Housing Education as a Local Mitigation Approach: seven case studies

In 1992, USAID-funded a 12-month study to investigate post-disaster housing education projects in the following countries: Jamaica; Dominica; St Vincent; Madagascar; Solomon Islands; Colombia; and Yemen. The case studies resulting from this project provide very illuminating critiques of these projects, and are of great value for the depth of their analysis.

Available from: J.Parker and Associates, 6527 Robin Rd, Dallas, Texas 75209-5320, USA.

IRDP Case Studies

A large number of case studies were commissioned as part of the International Relief/Development Project (IRDP). Eleven of these have been published in *Rising from the Ashes* (see section 1 of this compendium). The other case studies can be ordered seperately. The following are relevant to building improvement programmes in hazard-prone areas.

21. Armero, Colombia – volcano landslides
22. Tumaco – earthquake, floods
25. Ecuador – earthquake
26. Guatamala – earthquake
27. Mexico – earthquake
30. Philippines – typhoon
41. Popayon, Colombia (in Spanish) – earthquake

IRDP; 1989;
Available from: Disaster Management Center, University of Wisconsin, 432 North Lake St, Madison, WI 53706, USA.

Shelter After Disaster
Davis, I.

This contains much useful case study material from the 1970s, especially on building improvement projects following the Guatamala earthquake of 1976.

Oxford Polytechnic; 1978; 127pp; £10
Available from: Oxford Centre for Disaster Studies, PO Box 137, Oxford OX4 1BB, UK.
Also available in Spanish.

The Martyred City: death and rebirth in the Andes
Oliver-Smith, A.

Yungay, a town in Peru, was devasted by an earthquake and avalanche in 1970. The book provides a sympathetic account of the survivors efforts to rebuild their lives after the disasters. Many of the issues of post-disaster reconstruction are vividly brought out in this study, and many lessons can be learnt through the experiences of the Yungay community. This is one of the few detailed, long-term anthropological studies of a community following a disaster.

Waveland; 1992; 280pp; $9.95
Available from: Waveland Press Inc, PO Box 400, Prospect Heights, Illinois 60070, USA.

Power, Choice and Vulnerability: a case study in disaster mismanagement in south India 1977-88
Winchester, P.

The social and economic factors underlying vulnerability to natural disasters in south India are discussed in detail. The author is highly critical of government policies which have addressed vulnerability as a narrow technical problem.

James and James; 1992; 224pp; £29.50/$49.50
Available from: James and James Ltd, 5 Castle Rd, London NW1 8PR, UK.

Herramientos Para La Crisis: destastres, ecologismo y formacion profesional
Wilches-Chaux, G.

This comprehensive study is based primarily on the reconstruction programme undertaken by SENA following the Popayán earthquake in Colombia in 1983.

SENA; 1989; 250pp;
Available from: SENA, Calle 57 No. 8-69 Plazoleta, Apartado Aereo 53329, Bogota, Colombia.

Renovation of Low-income Housing in the Federal District: reconstruction of housing in the historic centre of Mexico City after the earthquakes of September 1985
UNCHS(HABITAT)

Following the earthquake of 1985, a large-scale government urban reconstruction programme was implemented: the Mexican Government's Programme to Renovate Low-income Housing. Through this programme, 48 800 dwellings were reconstructed in a 19-month period following the earthquake. This report discussess the social, technical and financial strategies of the programme.

UNCHS(HABITAT); 1989; 29pp; free
Available from: UNCHS(HABITAT), PO Box 30030, Nairobi, Kenya.
Also available in Spanish.

Improvement of low-cost housing in the Cook Islands to withstand tropical storms.
INTERTECT

Four detailed studies were carried out by INTERTECT on the impact of tropical storms and earthquakes on low-cost housing in the Pacific. One of these was on the Cook Islands. The other islands are Tonga (hurricanes and earthquakes), Fiji (hurricanes and earthquakes) and Tuvalu (hurricanes). Each report contains the following; an assessment of the risk of each island to natural hazards; an analysis of contemporary housing and housing trends; vulnerability analysis of low-cost housing; vulnerability reduction strategies; recommendations for comprehensive vulnerability reduction. The main drawback of these studies is that vulnerability is addressed as a narrow technical problem, and broader social factors affecting vulnerability are not considered (INTERTECT has also produced similar types of studies of low-cost housing on the Solomon Islands, Dominican Republic and Jamaica – see INTERTECT's publications catalogue).

INTERTECT; 1982; 72pp, 59pp, 75pp, 46pp; from $9.50 to $15 each
Available from: INTERTECT, 3511 North Hall St, Suite 302, Dallas, Texas 75219, USA.

Reducing Earthquake Losses in Rural Areas: a case study of eastern Turkey
Spence, R. and Coburn, A.

This is a technical study of the vulnerability of low-income housing to earthquakes in eastern Turkey and of possible methods of improving seismic-resistance of such buildings. The main chapters of the report are: buildings and earthquake vulnerability in eastern Anatolia; strengthening stone masonry buildings; tests on alternative improvement techniques; modelling earthquake losses; and conclusions and recommendations.

Martin Centre; 1987; 41pp
Available from: The Martin Centre, 6 Chaucer Rd, Cambridge CB2 2EB, UK.

Vernacular Housing in Seismic Zones of India
INTERTECT/University of New Mexico

A comprehensive analysis of low-cost housing in seismic zones of India was made for this report. Different building types are identified and their vulnerability to earthquakes assessed. Recommendations are given on how to improve such buildings to resist earthquakes.

INTERTECT; 1984; 205pp; $20
Available from: INTERTECT, 3511 North Hall St, Suite 302, Dallas, Texas 75219, USA.

Indigenous Building Techniques of Peru and their Potential for Improvement to Better Withstand Earthquakes
INTERTECT and Carnegie-Mellon University

This is a very detailed study of indigenous building techniques in rural Peru. The study identifies ways in which simple adaptions can be made to such technologies in order to make them more earthquake-resistant. Such methods of improving construction are illustrated with simple diagrams and pictures.

INTERTECT; 1981; 313pp; free
Available from: INTERTECT, 3511 North Hall St, Suite 302, Dallas, Texas 75219, USA.
Also available in Spanish.

12. Audio-visual materials

There is very little audio-visual material that is currently obtainable on building improvement programmes. While much material has been produced for previous projects, methods of disseminating this have not usually been established. Project workers will need to make enquiries from government departments, NGOs and research establishments to see whether any audio-visual material has been produced for previous projects in their own particular country.

Vision Habitat – Audio-Visual Catalogue
UNCHS(HABITAT)

A video/film catalogue can be ordered from any UNCHS(UNCHS(HABITAT)) office. None of the films have a specific focus on building improvement programmes in hazard-prone areas, but there are several films on co-operative housing and community participation.

Available from: Vision HABITAT, UNCHS(HABITAT), PO Box 30030, Nairobi, Kenya.

Building for Safety in Hazardous Areas
INTERTECT
15-minute video (Beta, VHS or 3/4); $22

A short film that animates the effects of earthquakes and high winds on buildings.

Available from: INTERTECT.
Also available in Spanish.

Disasters and Settlements: a series of four slide lectures
Davis, I.

The slide lectures are based on UNDRO's *Shelter After Disaster* manual.

Available from: UNCHS(HABITAT), PO Box 30030, Nairobi, Kenya.

Human Settlements and Disasters: five slide lectures
Davis, I. (ed)
£30 a lecture

Available from: Commonwealth Association of Architects, 66 Portland Place, London W1N 4AD.

Making Your House Safe Against Strong Winds
World Neighbors
Filmstrip; $10

This filmstrip, focusing on cyclone-prone areas of India, describes how to build a sturdy house from bamboo and mud.

Available from: World Neighbors, 4127 NW 122 St, Oklahoma City, OK 73120-8869, USA.

Communications
World Neighbors
Filmstrip; $30

This series of three filmstrips is intended to assist teachers of adult education with the use of audio-visual materials. The first filmstrip, *Portable Filmstrip/Slide Projectors*, discusses the making of filmstrips and use of portable projectors. The second filmstrip, *Planning Your Program*, explains how to use visual aids as part of a teaching programme. The third filmstrip, *Lifting*, explains how to make filmstrips from pictures in magazines.

Available from: World Neighbors, 4127 NW 122 St, Oklahoma City, OK 73120-8869, USA.
Also available in Spanish and Portuguese.

Ruben, Hero of Gandusari Village
World Neighbors
Filmstrip: $10

Using the example of introducing sanitation in an Indonesian village, this filmstrip highlights the need for consultation with, and the involvement of, the community when introducing new ideas.

Available from: World Neighbors, 4127 NW122 St, Oklahoma City, OK 73120-8869, USA.

At Home With Hurricanes
Building Research Establishment
Video; £35.76

Available from: Building Research Establishment, Garston, Watford, Hertfordshire WD2 7JR, UK.

In the Path of a Hurricane
PAHO
Video; $25

Available from: Preparedness and Disaster Relief Co-ordination Office, PAHO, Regional Office of WHO, 525 22nd St NW, Washington DC 20037, USA.

Myths and Realities of Natural Disasters
PAHO
Video; $25

Available from: Preparedness and Disaster Relief Co-ordination Office, PAHO, Regional Office of WHO, 525 22nd St NW, Washington DC 20037, USA.

The Earthquake in El Salvador
PAHO
Video; $50 (for four productions)

Available from: Preparedness and Disaster Relief Co-ordination Office, PAHO, Regional Office of WHO, 525 22nd St NW, Washington DC 20037, USA.

Disaster Chronicles No One: terremoto en popayan
PAHO
$25

This a report of the 1983 earthquake in 1983 in Popayon, based around 63 slides.

Available from: Preparedness and Disaster Relief Co-ordination Office, PAHO, Regional Office of WHO, 525 22nd St NW, Washington DC 20037, USA.

Disaster Chronicles No 2: earthquake in Mexico City
PAHO
$25

Available from: Preparedness and Disaster Relief Co-ordination Office, PAHO, Regional Office of WHO, 525 22nd St NW, Washington DC 20037, USA.

Portable Video Production
Video Tiers-Monde
Video; $250

Three workbooks and a video cassette are included in this training kit on how to produce videos.

Available from: Video Tiers-Monde, 3575 Saint-Laurent, Suite 608, Montreal, Quebec H2X 2T7, Canada or Instituto Para America Latina, Apartado Postal 270031, Lima 27, Perux.
Also available in Spanish, French and Italian.

13. General resource material

The following sources provide comprehensive bibliographies and directories of organizations.

HABITAT Directory
UNCHS(HABITAT)

The *HABITAT Directory* provides information on more than 1700 institutions and organizations throughout the world which are active in the field of human settlements.

UNCHS(HABITAT); 1986; 377pp; free
Available from: UNCHS(HABITAT), PO Box 30030, Nairobi, Kenya.
Available in English, French and Spanish.

Directory of Human Settlements Management and Development Training Institutions in Developing Countries
UNCHS(HABITAT)

This directory contains annotated entries of 149 institutions.

(UNCHS)HABITAT; 1991; 172pp; free
Available from: UNCHS(HABITAT), PO Box 30030, Nairobi, Kenya.
Available in English, French and Spanish.

Guides to Information Sources – 1. Building and Construction, 2. Housing, 3. Settlement Planning
UNCHS(HABITAT)

These guides were compiled specifically for the needs of developing countries. Each volume includes lists of national and international organizations, information on periodicals, books and reports, visual materials, conference proceedings, and other sources of information. This information has not been updated since 1980, but may still be useful.

UNCHS(HABITAT); 1980; 140, 102, 150pp; free
Available from: UNCHS(HABITAT), PO Box 30030, Nairobi, Kenya.
Available in English, French and Spanish.

Bibliographic Lists
UNDRO Library

Bibliographic lists can be requested from UNDRO library on a number of fields related to hazards and risks.
Available from: DHA-UNDRO, Palais des Nations, Geneva, Switzerland.

DESINDEX: Bibliografia Sobre Desastres (two volumes)
PAHO/WHO

The Centre de Documentacion de Desastre has compiled this extensive bibliography with more than 1600 annotated entries. The bibliography is written in Spanish, but contains a large number of English publications which are annotated in English. It contains both a subject index and author index.

PAHO; 1991/2;350pp and 175pp
Available from: Centro de Documentacion de Desastres, PAHO, Apart. Postal 3745-1000, San José, Costa Rica.

World Disasters Report 1993
International Federation of Red Cross and Red Crescent Societies

The report provides a detailed survey of the varied and complex issues concerning disasters worldwide with particular reference to the following: the regional distribution of disasters and the main issues relating to disaster response; trends in the changing nature of disasters; the causes and effects of today's disasters and the vulnerability of certain groups of people to such events; common questions raised and misconceptions held about disasters; the impact and effects of the main common types of disasters; various disaster definitions, developed to assist disaster analysis and research; a comprehensive compilation of global disaster statistics; and a full list of organizations involved in disasters worldwide.

IFRC; 1993; 126pp; US $49

Available from: Kluwer Academic Publishers, PO Box 322, 3300 AH Dordecht, The Netherlands and PO Box 358, Accord Station, Hingham, MA 02018-03587, USA.

Bibliographic Notes
UNCHS(HABITAT)

UNCHS(HABITAT) periodically issues bibliographic notes, each of which contain a carefully selected number of annotated entries. The following are currently available still: squatter settlements; transportation in human settlements; settlement planning; women and shelter; low-cost sanitation; urban management; rural settlements; resettlement in rural areas; community participation; small-scale production of building materials; women and human settlements; earth construction technology; co-operative housing; and solid-waste management.

Available from: UNCHS(HABITAT), PO Box 30030, Nairobi, Kenya.

14. Journals and newsletters

The primary basis for selection of journals and newsletters for this section has been that they have a practical focus and are issued free of charge, or with only a minimal subscription fee.

General disasters

DHA News (formerly *UNDRO News*)

A newsletter produced bimonthy by the United Nations Disaster Relief Office.

Available from: DHA-UNDRO, Palais des Nations, 1211 Geneva 10, Switzerland.

Stop Disasters: news from the International Decade for Natural Disaster Reduction

A newsletter for the International Decade for Natural Disaster Reduction. It is published bimonthly in English, French, Spanish and Italian.

Available from: IDNDR Secretariat, Palais des Nations, 1211 Geneva 10, Switzerland.

Disaster Preparedness in the Americas

The newsletter of the Emergency Preparedness and Disaster Relief Co-ordination Program of the Pan American Health Organization.

Available from: Disaster Preparedness in the Americas, PAHO, 525 23rd St NW, Washington, DC 20037, USA.

The Ark

Newsletter of the Pan-African Centre for Emergency Preparedness and Response.

Available from: UNECA, PO Box 3050, Addis Ababa, Ethiopia.

Red Cross/Red Crescent

Magazine of the International Federation of Red Cross and Red Crescent Societies.

Available from: IFRC, PO Box 372, CH 1211 Geneva 10, Switzerland.

Caribbean Disaster News

A newsletter covering disaster management in the Caribbean. Contains reports, safety tips, news briefs, and events diary.

Available from: Pan-Caribbean Disaster Perparedness and Prevention Project, PO Box 1399, St Johns, Antigua.

National Hazards Observer

A newsletter produced by the Natural Hazards Research and Applications Information Center, which exists to strengthen communications between researchers and organizations concerned with mitigating the effects of natural disasters. The *National Hazards Observer* is free to subscribers in the USA and $15 outside the USA.

Available from: Natural Hazards Research and Applications Information Center, Institute of Behavioural Science #6, Campus Box 482, University of Colorado, Boulder, Colorado 80309-0482, USA.

AODRO Newsletter

Newsletter of the Australian Overseas Disaster Response Organization. It is published quarterly.

Available from: Australian Overseas Disaster Response Organization, Level 1 491 Elizabeth St, Surry Hills NSW 2010, Australia.

Disasters: the journal of disaster studies and management

An academic journal covering a wide range of subjects within disaster management. Annual subscription is £79 institutions/£53 individuals. Single issue price is £23 institutions/£16.25 individuals.

Available from: Blackwells Publishers, 108 Cowley Rd, Oxford OX4 1JF, UK.

SECED Newsletter

A quarterly newsletter of the Society of Earthquake and Civil Engineering Dynamics.

Available from: SERCD Newsletter, Allott and Lomax, Fairbairn House, Ashton Lane, Sale, Manchester M33 1WP, UK.

NCEER Bulletin

A quarterly newsletter of the National Center for Earthquake Engineering Research.

Available from: NCEER, State University of New York at Buffalo, Red Jacket Quadrangle, Buffalo, NY 14261, USA.

Disaster Management

A quarterly newsletter focusing on all aspects of disaster management in India.

Available from: Joint Assistance Centre, H-65 South Extension Part 1, New Delhi 110049, India.

Natural Hazards: an international journal of hazards research and prevention

Available from: Kluwer Academic Publishers Group, Distribution Centre, PO Box 322, 3300 AH Dordrecht, The Netherlands.

Partnership News

The Bulletin of the Latin American Partnership to Enhance Co-operation in Earthquake Hazard Reduction.

Available from: Central United States Earthquake Consortium, 2630 East Holmes Rd, Memphis, TN 38118-8001, USA.

Low-cost housing and general development

HABITAT News

The official newsletter of UNCHS(HABITAT), which is published three times a year. It contains an inset called *NGO News* which serves as a forum for NGOs concerned with issues related to human settlements.

Available from: UNCHS(HABITAT), PO Box 30030, Nairobi, Kenya.

UNCHS(HABITAT) Shelter Bulletin

This bulletin contains recent news reports of housing projects and meetings.

Available from: UNCHS(HABITAT), PO Box 30030, Nairobi, Kenya.

Environment and Development Briefs – UNESCO

UNESCO Environment and Development Briefs are intended to improve the communication of scientific information to decision-makers. Issue number five is concerned with disaster reduction, and contains a useful introduction to the major issues involved.

Available from: UNESCO, 7 Place de Fountenoy, 75352 Paris 07 SP, France.

Caribbean Cyclone-resistant Housing Project Newsletter

This newsletter reports on the progress of the Caribbean Cyclone-resistant Housing Project.

Available from: Caribbean Cyclone-resistant Housing Project, University of West Indies, St Augustine, Trinidad and Tobago, West Indies.

CRDC Crane

The newletter of the Construction and Resource Development Centre, Jamaica.

Available from: CRDC, 11 Lady Musgrave Avenue, Kingston 10, Jamaica, West Indies.

Appropriate Technology

A quarterly magazine produced by Intermediate Technology Publications. It contains reports from development projects, the latest Appropriate Technology applications, news for ITDG, book reviews and a development diary. Annual subscription £18 institutions/£14 individuals.

Available from: Intermediate Technology Publications, 103-105 Southampton Row, London WC1B 4HH, UK.

Open House International

A quarterly journal concerned with housing and planning in developing countries. Annual subscription £45 institutions/ £30 individuals. Single and back issues can be ordered from the same address at £10 an issue. Vol 12 No 3, 1987, is a special issue on 'Homelessness and Disaster Response' containing 10 useful papers on housing and disasters.

Available from: Centre for Architectural Research and Development Overseas, School of Architecture, University of Newcastle, Newcastle upon Tyne, NE1 7RU, UK.

BASIN News

An international newsletter of BASIN – Building Advisory Service and Information Network. It contains four seperate sections on wall building, cementitious binders, roofing and earth building.

Available from: SKAT, Varnbueelstrasse 14, CH 9000, St Gallen, Switzerland.

LINKS

The quarterly newsletter of the Antigua Programme Centre of the Caribbean Conference of Churches. The Centre is concerned with disaster relief and shelter programmes in the eastern Caribbean.

Available from: Caribbean Conference of Churches, PO Box 616, Bridgetown, Barbados, West Indies.

Environment and Urbanization

Published twice yearly by the International Institute for Environment and Development, this journal aims to provide an effective means of exchange of ideas and information in the fields of human settlements and the environment. Previous issues include: *"Community-based Organizations"* (Vol 2 No 1, 1990) . Subscription £8 developing countries, £17 elsewhere. Back issues £5 developing countries, £8 elsewhere. Southern NGOs and teaching organizations can apply for free subscription.

Available from: IIED, 3 Endsleigh St, London WC1H 0DD, UK or IIED-América Latina, Piso 6, Cuerpo A, Corrientes 2835, 1193 Buenos Aires, Argentina.

Waterlines
Intermediate Technology Publications

Waterlines is a magazine on low-cost water and sanitation. It is aimed at project managers, policy-makers, trainers, administrators, and engineers. Back issues of *Waterlines* can be ordered through Intermediate Technology Publications. Cost: £3.50 an issue.

Available from: Intermediate Technology Publications, 103/105 Southampton Rd, London, WC1B 4HH, UK.

Footsteps
Tear Fund

Footsteps is a quarterly newsletter linking health and development workers. Each issue has a specific focus and attempts to introduce new ideas. It is distributed free of charge.

Available from: The Editor, 83 Market Place, South Cave, Brough, North Humberside HU15 2AS, UK.

15. Directory of organizations

The following organizations have been included on the grounds that they have developed some expertise relevant to building improvement programmes in hazard-prone areas. The subjects covered by these organizations include: disaster-resistant construction; low-cost housing; general disaster management; programme management; training for development workers. This directory is by no means exhaustive, but its aim is to identify key organizations in various areas of technical expertise and regional specialization.

UN and regional organizations

Department of Humanitarian Affairs
Palais des Nations
1211 Geneva 10
Switzerland

DHA (formerly UNDRO) is the focal UN agency for co-ordinating disaster relief and providing technical expertise.

United Nations Centre for Human Settlements UNCHS(HABITAT)
PO Box 30030
Nairobi
Kenya
Tel +254 2 333930, 520600, 520320
Fax +254 2 520724

UNCHS(HABITAT) is the UN agency concerned with human settlements. In this capacity it funds a wide range of projects, advises and assists national governments, and acts as a centre of knowledge and experience for human settlements development and management. It has also established, with support from DANIDA, a special training programme on community participation. UNCHS(HABITAT) publishes a wide range of publications, especially on low-cost housing, which can usually be ordered free of charge. Audio-visual materials are also produced by Vision Habitat, a section within UNCHS, and these can be hired from UNCHS offices. A publications list, Vision Habitat list and more details of expertise and services available through UNCHS can be requested from UNCHS's head office in Nairobi, or from any of the UNCHS regional offices:

- UNCHS Amman Office, PO Box 35286 Amman, Jordan
- UNCHS Bangkok Office, Room 1013, Block B, Rajadamnern Avenue, Bangkok 10020, Thailand
- UNCHS Budapest Office (for eastern Europe), H 1400 Budapest, PF 83, Hungary;
- UNCHS Geneva Office, Palais des Nations 1211 Geneva 10, Switzerland
- UNCHS New York Office, DC2, Room 946, UN, New York 10017, USA
- UNCHS Mexico City Office, c/o CEPAL – Naciones Unidas, Apartado Postal 6-718, Mexico City, Mexico, and
- UNCHS Ottawa Office (for North America and the Caribbean), Suite 147, 130 Albert St, Ottawa, Canada K1P 5G4.

United Nations Development Program (UNDP)
1 UN Plaza
New York, NY 10017
USA

UNDP resident representatives act on behalf of DHA-UNDRO in the countries to which they are assigned and engage in disaster preparedness, mitigation and relief activities.

United Nations High Commissioner for Refugees (UNHCR)
Palais des Nations
CH-1211 Geneva 10
Switzerland

United Nations Education, Science and Cultural Organization (UNESCO)
7 Place de Fontenoy
75700 Paris
France
Tel +33 1 45 68 10 00
Fax +33 1 45 67 16 90

UNESCO – Regional Office for Asia and the Pacific
Box 967 Prakanong Post Office
Bangkok 10110
Thailand
Tel +66 391 0577
Fax +66 391 0866

This regional office of UNESCO has funded several major projects on earthquake and cyclone-resistant school buildings.

UNDP/UNIDO Regional Network in Asia for Low-cost Building Materials Technology and Construction Systems
Office of Regional Secretariat
10th Floor, Allied Bank Building
Ayala Avenue
Makati
Metro Manila
Philippines

Preparedness and Disaster Relief Co-ordination Office
Pan American Health Authority (PAHO)
525 22nd St NW
Washington DC 20037
USA

PAHO has produced a large amount of written and visual materials on disaster management. This includes: the quarterly newsletter *Disaster Preparedness in the Americas*; a booklet of abstracts of publications and visual materials on disaster related issues; and *Disaster Mitigation Guidelines for Hospitals and Other Health Care Facilities in the Caribbean*.

International Labour Office
CH 1211
Geneva 22
Switzerland
Tel +41 22 799 61 11
Fax +41 22 798 63 58

A major part of ILO's work consists in the provision of advice and technical assistance to individual countries. Much of this activity lies in such fields as training, employment promotion, and the development of co-operatives. ILO publishes a large number of documents, including several training manuals on co-operative management. ILO has offices in many countries through which publications can be ordered.

International Federation of Red Cross and RedCrescent Societies
Policy Department
17 Chemin des Crets
Petit Saconnex
CH 1211
Geneva 19
Switzerland
Tel +41 22 7304449
Fax +41 22 7330395

Organization of American States
Dept of Regional Development and Environment
17th St
NW Washington
DC 20006
USA

La Red

This is a recently formed Latin American network for research, information and training on disaster prevention and management. The following organizations are currently in La Red: ITDG, Peru; FLASCO, Costa Rica; COMECSO, Mexico; Disaster Research Unit, Canada; CED, Brasil; OSSO, Colombia; Partners of the Americas, Ecuador; FUNCOP, Colombia; ONAD Colombia; and CIPER, Costa Rica.

Australia

Cyclone Structural Testing Station
James Cook University of Queensland
Townsville
North Queensville 4811
Australia
Tel +61 (0) 77 81 4754
Fax +61 (0) 77 75 1184

The Cyclone Structural Testing Station is engaged in the field of structural testing and design against the effects of high winds on low rise buildings. It undertakes research consultancy and information dissemination work. It is primarily concerned with cyclone-resistant buildings in Australia and the Pacific region. Information bulletins and technical reports can be ordered from the above address.

Australian Overseas Disaster Response Organization (AODRO)
PO Box K425
Haymarket 2000
Australia

The Australian Overseas Disaster Response Organization co-ordinates overseas disaster assistance on behalf of Australian NGOs and the general community. An important aspect of its work is disaster preparedness training for NGOs in the Pacific Islands. This includes the promotion of disaster-resistant vernacular housing through its workshops and publications.

Bangladesh

PACT Bangladesh
78, Satmasjid Rd
Dhanmondi
Dhaka
Bangladesh

Brazil

Centre for Disaster Studies (CED)
Federal University of Paraiba
Rua Manoel Barros de Oliveira 327
Barrio Universitario
58,108-125 Campina Grande
Paraiba
Brazil
Tel +55 83 333 1833
Fax +55 83 333 1945

CED has a multidisciplinary team carries out research various issues relating to natural disasters, including vulnerability studies.

Canada

Disaster Research Unit
University of Manitoba
Dept of Anthropology
Winnepeg
R3T 2N2 Manitoba
Canada
Tel +1 204 474 8999
Fax +1 204 275 0846

The Disaster Research Unit undertakes research and dissemination activities related to disasters.

The Caribbean

Construction Resource and Development Centre (CRDC)
11 Lady Musgrave Avenue
Kingston 10
Jamaica
West Indies
Tel +1 809 978 3945 and 978 4061
Fax +1 809 978 3945

The Construction Resource and Development Centre is a NGO concerned with shelter issues in Jamaica and The Caribbean. It has worked with a wide range of government NGOs and academic institutions and has experience in the following areas: builders' training; disaster mitigation; low-income housing design; the environment. CRDC is especially concerned with the informal housing sector and is in the process of setting up a resource centre for the informal housing sector in order to support the production of housing solutions in this sector.

Caribbean Cyclone-resistant Housing Project
The University of West Indies
St Augustine
Trinidad and Tobago

West Indies
Tel +1 809663 2046 Ext 3434
Fax +1 809662 4414

The Caribbean Cyclone-resistant Housing Project is a joint research project being undertaken by the University of West Indies and the University of Waterloo, Canada and funded by the International Development Research Centre, Canada. Its objective is to improve, through research, cyclone-resistant construction techniques and practice in lower-income housing projects in the CARICOM region. A key element within the project is to disseminate the research results on a widescale in the region. An information bulletin is produced by the project and can be obtained from the above address.

Caribbean Conference of Churches
PO Box 616
Bridgetown
Barbados
West Indies
Tel +1 809 427 2681
Fax +1 809 429 2075

The Caribbean Conference of Churches is an umbrella organization for the ecumenical movement in The Caribbean. It is represented in 34 countries of the English, French, Dutch and Spanish speaking Caribbean. One of the interests of the CCC is in disaster relief and reconstruction. It has produced the manual *When You Build a House*, which is a manual of construction details for Caribbean houses with emphasis on protection from strong winds.

China

China Building Technology Development Centre
19 Che Gong Zhuang St
Beijing
China
Tel +86 1 8992613

The China Building Technology Development Centre is an institution under the Ministry of Urban and Rural Construction and Environmental Protection. Its purpose is to promote and develop building technology and urban and rural construction. It contains four institutes and six divisions including the Institute of Rural Development and the Division of Audio and Video Recording.

Beijing Institute of Architectural Design and Research
62 South Lishi Rd
100045
Beijing
China
Tel +86 1 8512255/233
Fax +86 1 8034041

This institute specializes in civil architectural design and has conducted research on the resistance to earthquakes of various forms of building structure, both engineered and non-engineered, including brick masonry and reinforced concrete.

Colombia

Asociacion Colombiana de Ingenieria Sismica (AIS)
Avenida 15 No 118-03 Local 122
Apartado 092416
Bogota
Colombia
Tel +57 1 215 2406

The Colombian Association of Earthquake Engineering.

Office Nacional Para La Prevencion y Atencion De Desastres (ONAD)
(National Office for Disaster Prevention and Attention)
Calle 7 No 6-54 Piso 3
Bogota
Colombia
Tel +57 1 283 4966

ONAD is a governmental agency dedicated to the co-ordination and promotion of disaster management in Colombia, at local, regional and national level.

Federacion De Asociaciones De Vivienda Popular (FEDEVIVIENDA)
Avenida Calle 40 No 15-69
Apartado 57059
Bogota
Colombia
Tel +57 1 288 0711

The Federation of Low-cost Housing Associations in Colombia

Fundacopm Servocio De Vivienda Popular (SERVIVIENDA)
Calle 48 No 14-61
Bogota
Colombia
Tel +57 1 287 9666

The Foundation for Low-cost Housing Services.

Servicio Nacional De Aprendizage (SENA)
Calle 57 No 8-69 Plazoleta
Apartado 53329
Bogota
Colombia
Tel +57 1 217 0177

The government's National Learning Service which has had considerable involvement in post-earthquake reconstruction programmes.

Fundacion Para La Comunication Popular (FUNCOP)
Calle 5 No 877
PO Box 2096 Apartado Aereo 1280
Popayan
Colombia
Tel +57 282 41541,44806
Fax +57 282 40409

FUNCOP is an NGO dedicated to the promotion and strengthening of grassroots organizations in Cauca region. Disaster issues are incorporated into its programmes.

Seismological Observatory of the Southwest (OSSO)
Ciudad Universitaria Meléndez
Torre de Ingenierí piso 3
PO Box 25-360
Cali
Colombia
Tel +57 23 397 222
Fax +57 23 313 418

OSSO has a bibliographical data bank on disasters, and a publications programme, and specializes in seismological research.

Costa Rica

FLACSO
Calle 29, Avienda 9, Casa 942
Apartado Postal 5429
1000 San José
Costa Rica
Tel +506 57 0533/34 9868
Fax +506 21 5671

This is an intergovernmental institution dedicated to the promotion and dissemination of social science research. It is involved in studies of the responses to the Costa Rica earthquake and on urban community vulnerability and response to disasters in Central America.

Church Council for Emergencies and Reconstruction (CIPER)
PO Box 62
1001 San José
Costa Rica
Tel +506 25 1249
Fax +506 25 1249

CIPER is a NGO which undertakes training, research, and community assistance in relation to disasters.

France

CRATerre
Centre Simone Signoret
BP 53F - 38090
Ville Fontaine
France
Tel +33 74 96 60 56
Fax +33 74 96 04 63

CRATerre is the International Centre for Building with Earth. It conducts research and development documentation and consultancy on the promotion of earth as a building material. It also conducts intensive training courses in earth construction in France and overseas. For more information on its activities, training courses and publications write to the address given.

Germany

Building Advisory Service and Information Network (BASIN)
GTZ (GATE)
Transport and Construction Division
Postfach 5180D 6236
Eschborn
Germany
Tel +49 (0) 6196 793130
Fax +49 (0) 6196 791115

BASIN was set up in 1989 on behalf of the German Appropriate Technology Exchange (GATE), in collaboration with Intermediate Technology (UK) SKAT (Switzerland) and CRATerre (France), to provide information on appropriate planning, construction and production of building materials using local resources and the promotion of the local building industry in developing countries. BASIN also offers an advisory service and training courses for professionals involved in construction projects in developing countries. Further information and a list of publications available through BASIN can be obtained from the above address.

India

Central Building Research Institute
Roorkee - 247 667
Undra Pradesh
India
Tel +91 (0) 1332 2243 2428 2293
Fax +91 (0) 1332 2272

This is a leading Asian research institute with expertise in a very wide range of areas from highly technical research to low-cost housing issues. Major areas of research include: disaster-resistant housing (floods/cyclones); low-cost alternative building material and components; rural buildings and environment; project management; and training and extension work. More information on the research and activities of the institute and a list of its publications can be requested from the address given.

Dept of Earthquake Engineering
University of Roorkee
Roorkee 247 667
India
Tel +91 Roorkee 2349, 2405, 2604

Housing and Urban Development Co-operation (HUDCO)
HUDCO Office
Lodhi Rd
New Delhi 110 003
India
Fax +91 11 4625308

Provides loans for reconstruction of dwellings.

Structural Engineering Research Centre
Council of Scientific and Industrial Research
TTTI
Taramani PO Box Madras 600 113
India
Tel +91 44 2352122/ 2352211
Fax +91 44 2350508

SERC has expertise in the following: wind effects on buildings; surveys of post-disaster structural damage due to cyclones; risk analysis for cyclones; cyclone wind map of coastal regions of India; guidelines on cyclone-resistant structures; international workshops on cyclone disaster mitigation.

UNNAYAN
36/1A Garcha Rd
Calcutta 700 019
India

An NGO with considerable experience in community-based reconstruction programmes.

Habitat Technology Network – A Video-based Technology Documentation on Alternative Methodologies of Building
Building Centre
Sarai Kale Khan
East Nizamuddin
New Dehli 110013
India
Tel +91 11 617460/698188

The Habitat Technology Network aims to provide information and data on low-cost building techniques through regular issues of a series videos. The intention is to disseminate the technologies to a wide number of people involved in low-cost construction. The videos serve as a visual newsletter. Twelve videos are produced each year at a cost of Rs500 each.

Centre for Disaster Management and Rural Reconstruction
National Institute of Rural Development
Rajendranagar
Hyderabad 500 030
Andhra Pradesh
India
Tel +91 (0) 842 245270
Fax +91 (0) 842 245277

The CDM and RR undertakes training, research and policy analysis on disaster management issues in India.

Mexico

Sociedad Mexican de Ingenieria Sismica AC
Camino Santa Teresa 187
Apartado Postal 70 227
Mexico DF 04510

This is an information and advisory centre for earthquake-resistant construction.

The Mexican Social Sciences Council (COMECSO)
Torre II de Humanidades
8 Piso, Ciudad Universitaria
04510 México DF
México
Tel +52 5 623-0210/623-0209
Fax +52 5 548-4315

COMESCO has one working group dedicated to the social study of disasters, and undertakes teaching of disaster studies.

CENAPRED — Centro Nacional de Prevención de Desastres
Av Delfin Madrigal No 665
Col Santo Domingo
Coyoacan CP 04360
Mexico DF
Tel +52 5 606 9520
Fax +52 5 606 1608

The technical agency of the Mexican Government responsible for researching, developing and co-ordinating the implementation of technologies for disaster prevention and mitigation, and for the provision of training.

The Pacific

The University of the South Pacific
Department of Technology
PO Box 1168
Suva
Fiji
Tel +679 314900

Currently working on project on the study of appropriate building materials and methods for cyclone-resistant building in the South Pacific region.

Peru

CRATerre America Latina
Apartado Postal 5603
Correo Central
Lima 1

This specializes in earthquake-resistant earth construction.

ITDG
Av Jorge Chávez 275
Casilla 18-0620
Lima 18 (Miraflores)
Peru
Tel +51 14 46 6621/46 7324
Fax +51 14 46 6621

ITDG Peru has developed a special expertise in disaster issues, including the implementation of disaster prevention, mitigation and preparedness projects.

The Philippines

Philippine Institute of Volcanology and Seismology
5th and 6th Floors
Hizon Building
29 Quezon Avenue
Quezon City
Tel +63 712 61 10/11/12/13/14/15

The government agency responsible for monitoring and studying volcanoes and earthquakes in the Philippines and making recommendations for preparedness and mitigation. It has produced a large number of publications.

Switzerland

Schweizerische Kontakstelle fur Angepassete Technik (SKAT)
Tigerbergstrasse 2
CH 9000 St Gallen
Switzerland
Tel +41 (0) 71 30 25 85
Fax +41 (0) 71 22 46 56

SKAT (Swiss Centre for Appropriate Technology) is a documentation centre and consultancy group which is engaged in promoting appropriate technologies in developing countries. One of SKAT's main activities is in building materials. It has special expertise in roof materials and roof construction and has been running the Fibre-concrete Roofing Advisory Service.

Thailand

Asian Disaster Preparedness Centre
Asian Institute of Technology
PO Box 2754
Bangkok 10501
Thailand
Tel +66 2 516 0110
Fax. +66 2 516 2126

This is the major disaster training and research centre for South-east Asia.

Turkey

Earthquake Engineering Research Institute
Middle East Technical University
06531 Ankara
Turkey

United Kingdom

Building Research Establishment
Garston
Watford
Hertfordshire WD2 7JR
UK
Tel +44 (0) 923 894040
Fax +44 (0) 923 664010

The BRE is a government research laboratory which specializes in the technical aspects of buildings, fire prevention and related environmental issues. It provides an advisory service for the construction industry, a technical consultancy service and has an overseas division. The BRE has, in the past, given considerable attention to low-cost hazard-resistant construction although this has now reduced. Among its many publications is a series called *Overseas Building Notes and Information Papers* which contains, among other things, booklets on building to resist earthquakes, strong winds and production of building materials in developing countries. BRE has also produced a list of building research centres in developing countries although this has not been updated since 1978

(*Overseas Building Notes 163*, £2). Further information on BRE services and publications can be obtained by writing to the BRE Technical Consultancy at the address given.

Cranfield Disaster Preparedness Centre
Royal Military College of Science
Shrivenham
Swindon
Tel +44 (0) 793 785287
Fax +44 (0) 793 782179

Organizes disaster management training workshops both in UK and in-country and has particular expertise in logistics, drought relief and mitigation.

Cambridge Architectural Research
The Eden Centre
47 City Rd
Cambridge
CB1 1DP
Tel +44 (0) 223 460475
Fax +44 (0) 223 464142

Specialist experience in low-cost, disaster-resistant housing and in disaster mitigation.

Intermediate Technology Development Group
Myson House
Railway Terrace
Rugby CV21 3HT
UK
Tel +44 (0) 788 560631
Fax +44 (0) 788 540270

Intermediate Technology offers expertise in a wide range of technical areas including construction where it has developed particular specialization in cements and binders. Its publishing arm, Intermediate Technology Publications produces a *Books by Post* catalogue, offering a very comprehensive selection of publications in development and intermediate technology and a scheme for purchasing books in local currency. (Address: 103-105 Southampton Row, London WC1B 4HH, UK Tel +44 (0)71 436 9761, Fax +44 (0)71 436 2013).

International Institute for Environment and Development
3 Endsleigh St
London WC1H 0DD
UK
Tel +44 (0) 71 388 2117
Fax +44 (0) 71 388 2826

IIED is an independent, non-profit-making organization which promotes sustainable patterns for development in developing countries. It has research programmes in a number of areas including a human settlements programme. The human settlements programme includes research on community-level initiatives in low-cost housing and services and aims to promote techniques for community-based action. The programme has been based on collaborative work with institutions and NGOs in developing countries and has an important networking role. IIED has a Latin American office at the following address: IIED-America Latina, Piso 6, Cuerpo A, Corrientes 2835, 1193 Buenos Aires, Argentina, Tel +54 1 961 3050, Fax +54 1 961 1854.

International NGO Training and Research Centre (INTRAC)
PO Box 563
Oxford
OX2 6RZ
Tel +44 (0) 865 201851
Fax +44 (0) 865 201852

INTRAC provides training, research and consultancy support to the NGO community, offering a wide range of training courses. Its particular areas of expertise are: NGO management; income generation; monitoring and evaluation.

Oxford Centre for Disaster Studies
PO Box 137
Oxford
OX4 1BB
Tel +44 (0) 865 202772
Fax +44 (0) 865 202848

This new centre draws on more than 20 years of experience in training, research and consultancy on disaster management in developing countries, covering in particular risk assessment, preparedness planning, mitigation strategies, relief management, and reconstruction planning.

Teaching Aids at Low Cost (TALC)
PO Box 49
St Albans
Hertfordshire AL1 4AX
UK
Tel +44 (0) 727 53869
Fax +44 (0) 727 46852

TALC is an organization concerned with providing low-cost teaching aids for developing countries, primarily in the field of primary health care. A list of books and visual materials that can be ordered through TALC can be obtained from the above address.

United States of America

National Center for Earthquake Engineering Research (NCEER)
University of New York at Buffalo
Red Jacket Quadrangle
Buffalo
New York 14261
USA
Tel +1 716 636 3391
Fax +1 716 636 3399

This institution undertakes highly technical research for earthquake-resistant engineered structures in the USA. It has established an information service which has compiled a huge database on earthquake engineering, natural disasters and related subjects. Bibliographic lists on a wide range of subjects can be ordered from the information service.

International Technical Consultants for Emergency Management (INTERTECT)
3511 North Hall St
Suite 302
Dallas
Texas 75219
USA
Tel +1 214 521 8920
Fax +1 214522 9332

INTERTECT is a private consultancy which has extensive experience of assisting UN agencies, governments and NGOs in disaster management. This includes low-cost disaster-resistant housing. INTERTECT has been involved in numerous post-disaster housing programmes and has assisted in the production of several training manuals for builders in disaster-resistant construction. These are among the many publications produced by INTERTECT. A publication list and more information on the services that INTERTECT can offer may be obtained from the above address.

INTERWORKS
116 North Few St
Madison
Wisconsin 53703
USA
Tel +1 608 251 9440
Fax +1 608 251 9150

This organization is concerned with the development of training in disaster management. It is currently co-ordinating the UNDP/UNDRO Disaster Management Training Programme.

Natural Hazards Research and Applications Information Centre
Institute of Behavioral Science #6
Campus Box 482
University of Colorado
Boulder
Colorado 80309-0482
USA

Bibliographies can be obtained from the above address.

Office of Foreign Disaster Assistance
USAID
Washington DC
USA

OFDA produces a number of publications, including disaster profiles of most developing countries.

Venzuela

SOCVIS - Sociedad Venzolana de Ingenieria Sismica
Av Mohendan #9
La Castellana
CARACAS
Venzuala

This is primarily oriented towards civil engineers. It is also a contact organization for the Latin American and Caribbean Network Centers on Earthquake Engineering.